T0338003

Introduction to Numerical Methods for Time Dependent Differential Equations

Introduction to Numerical Methods for Time Dependent Differential Equations

Heinz-Otto Kreiss
Träskö-Storö Institute of Mathematics
Stockholm, Sweden

Omar Eduardo Ortiz
Facultad de Matmática Astronmía y Física
Universidad Nacional de Córdoba
Córdoba, Argentina

Library of Congress Cataloging-in-Publication Data:

Kreiss, H. (Heinz-Otto)
    Introduction to numerical methods for time dependent differential equations / Heinz-Otto Kreiss, Träskö-Storö Institute of Mathematics, Stockholm, Sweden, Omar Eduardo Ortiz, Facultad de Matmática Astronmía y Física, Universidad Nacional de Córdoba, Córdoba, Argentina.
        pages cm
    Includes bibliographical references and index.
    ISBN 978-1-118-83895-2 (cloth)
    1. Diffferential equations, Partial—Numerical solutions. I. Ortiz, Omar Eduardo, 1965– II. Title.
    QA374.K915 2014
    515'.353dc23   2013042036

Printed in the United States of America.

10 9 8 7 6 5 4 3 2 1

To our families

CONTENTS

PREFACE

This book is based on the class notes of a course that H. Kreiss taught in the Department of Mathematics at UCLA in the year 1998. The original notes were then used by many other people. In particular, O. Ortiz used those original notes in a course taught in Fa.M.A.F., Universidad Nacional de Córdoba, in 2007 and 2010. The positive feedback from students taking these courses encouraged us to write the book.

Our intention was always to write a short book, suitable for an introductory, self-contained course that places emphasis on the fundamentals of time dependent differential equations and their relation to the numerical solutions of these equations.

The book is divided into two parts. The first part, from Chapter 1 to Chapter 6, deals with ordinary differential equations (ODEs) and their approximations. Chapter 1 is a simple presentation of the fundamental ideas in the theory of scalar equations. Chapter 2 is the core of the first part of the book, where most of the important concepts on finite-difference approximations are introduced and explained for the most basic method of all, the explicit Euler method. The remaining chapters deal with higher-order approximations, implicit methods, multistep methods, and systems of ODEs.

Our intention in this book is to emphasize the principles on which the theory is based. This is, one first needs to understand clearly the theory of scalar ordinary differential equations with constant coefficients. Then the variable coefficient problems are approached by appealing to the principle of frozen coefficients, which allows one to split the variable coefficient problem into many constant coefficient problems. Nonlinear problems are treated via the principle of linearization, which turns a nonlinear problem into a linear variable coefficient problem, which then decomposes into constant coefficient problems via the principle of frozen coefficients. For systems of ordinary differential equations we require that we can diagonalize the system, and then we just need to understand scalar equations.

The second part of the book deals with partial differential equations in one space dimension and their approximations. The basics of Fourier series and interpolation are presented in Chapter 7. Chapters 8, 9 and 10 are devoted to the concepts of well-posedness and numerical approximations for both Cauchy problems and initial boundary value problems. We start the discussion by treating in detail three basic equations: the one-way wave equation (or advection equation), the heat equation, and the wave equation. In Chapter 11, the final chapter, we develop the idea of "when"and "why" nonlinear differential problems can be thought of as perturbation of a numerically computed solution, thus making the approximations meaningful.

We want to make clear that one first needs to understand the theory of differential equations, including estimates of the solution, after which one can prove the stability of the difference approximations by similar estimates. Therefore, the usual way is that one gets from the theory existence during a finite time, then one approximates the problem by difference approximations and computes the solution for as long as the approximation remains stable.

Exercises, with most of their solutions provided in an appendix, were included based on the conviction that solving exercises and computing are essential to the learning process of this subject.

All the software used in the preparation of the manuscript is open-source software run under GNU-Linux. The typeseting was done in LaTeX. The numerical computations for examples and exercises were written in C and compiled with gcc. The plots were generated using Gnuplot and Gimp.

HEINZ-O. KIRESS AND OMAR E. ORTIZ

Córdoba
April 2013.

ACKNOWLEDGMENTS

We would like to thank our students and colleagues for encouraging us to write this book, and other people for various contributions and help: in particular, Jarred Tanner for providing several solutions to exercises in the original Kreiss notes, Barbro Kreiss for being of invaluable help, and Gunilla Kreiss for carefully reviewing the manuscript and making suggestions.

This book was written partially with the support of grants 05/B454 from SeCyT, Universidad Nacional de Córdoba, and the Partner Group grant of the Max Planck Institute for Gravitational Physics, Albert-Einstein-Institute (Germany).

H.-O.K. and O.E.O.

PART I

ORDINARY DIFFERENTIAL EQUATIONS AND THEIR APPROXIMATIONS

CHAPTER 1

FIRST-ORDER SCALAR EQUATIONS

In this chapter we study the basic properties of first-order scalar ordinary differential equations and their solutions. The first and larger part is devoted to linear equations and various of their basic properties, such as the principle of superposition, Duhamel's principle, and the concept of stability. In the second part we study briefly nonlinear scalar equations, emphasizing the new behaviors that emerge, and introduce the very useful technique known as the principle of linearization. The scalar equations and their properties are crucial to an understanding of the behavior of more general differential equations.

1.1 Constant coefficient linear equations

Consider a complex function y of a real variable t. One of the simplest differential equations that y can obey is given by

$$\frac{dy}{dt} = \lambda y, \tag{1.1}$$

Introduction to Numerical Methods for Time Dependent Differential Equations, First Edition.
By Heinz-O. Kreiss and Omar E. Ortiz. Copyright © 2014 John Wiley & Sons, Inc.

where $\lambda \in \mathbb{C}$ is constant. We want to solve the initial value problem, that is, we want to determine a solution for $t \geq 0$ with given initial value

$$y(0) = y_0 \in \mathbb{C}. \tag{1.2}$$

Clearly,

$$y(t) = e^{\lambda t} y_0 \tag{1.3}$$

is the solution of (1.1), (1.2). Let us discuss the solution under different assumptions for the λ constant. In Figures 1.1 and 1.2 we illustrate the solution for $y_0 = 1 + 0.4i$ and different values of λ.

1. $\lambda \in \mathbb{R}, \lambda < 0$. In this case both the real and imaginary parts of the solution decay exponentially. If $|\lambda| \gg 1$, the decay is very rapid.

2. $\lambda \in \mathbb{R}, \lambda > 0$. The solution grows exponentially. The growth is slow if $|\lambda| \ll 1$. For example, for $\lambda = 0.01$ we have, by Taylor expansion,

$$e^{0.01t} = 1 + 0.01t + 0.0001t^2 + \cdots .$$

On the other hand, if $\lambda \gg 1$, the solution grows very rapidly.

3. $\lambda = i\xi, \xi \in \mathbb{R}$. In this case, the *amplitude* $|y(t)|$ of the solution is constant in time,

$$|y(t)| = |e^{i\xi t}||y_0| = |y_0|.$$

If the complex initial data y_0 is written as

$$y_0 = y_{0R} + iy_{0I}, \quad y_{0R}, y_{0I} \in \mathbb{R},$$

the solution is

$$\begin{aligned}
y(t) &= e^{i\xi t} y_0 \\
&= \big(\cos(\xi t) + i\sin(\xi t)\big)\big(y_{0R} + iy_{0I}\big) \\
&= \big(y_{0R}\cos(\xi t) - y_{0I}\sin(\xi t)\big) + i\big(y_{0R}\sin(\xi t) + y_{0I}\cos(\xi t)\big),
\end{aligned}$$

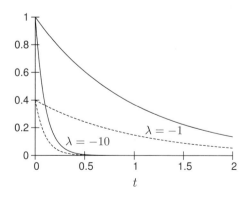

Figure 1.1   Exponentially decaying solutions. $\mathrm{Re}\,y$ shown as solid lines and $\mathrm{Im}\,y$ as dashed lines.

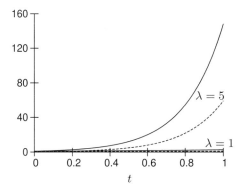

Figure 1.2   Exponentially growing solutions. $\mathrm{Re}\,y$ is shown as solid line and $\mathrm{Im}\,y$ as a dashed line.

which defines the real part $y_R(t)$ and imaginary part $y_I(t)$ of the solution. Both parts are oscillatory functions of t. The solution is highly oscillatory if $|\xi| \gg 1$. Figure 1.3 shows the solution for $\lambda = 2i$ and $y_0 = 1 + 0.4i$. Another representation of the solution is obtained if we write the initial data in amplitude-phase form,

$$y_0 = e^{i\alpha}|y_0|, \quad -\pi < \alpha \leq \pi.$$

One calls the modulus $|y_0|$ the *amplitude* of y_0 and the principal argument α the *phase* of y_0. The solution becomes

$$y(t) = e^{i\xi t}y_0 = e^{i(\xi t + \alpha)}|y_0| = \big(\cos(\xi t + \alpha) + i\sin(\xi t + \alpha)\big)|y_0|. \quad (1.4)$$

The real part of the solution with $\lambda = i$, $|y_0| = 1$, and $\alpha = -\pi/4$ is shown in Figure 1.4.

$$\mathrm{Re}\,y = \cos(t - \pi/4)$$

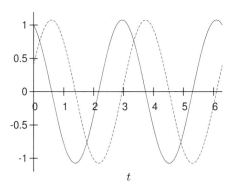

Figure 1.3   Oscillatory solution. $\mathrm{Re}\,y$ is shown as a solid line and $\mathrm{Im}\,y$ as a dashed line.

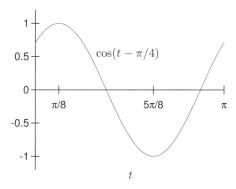

Figure 1.4   Real part of the solution (1.4).

4. *The general case.* Let

$$\lambda = \eta + i\xi, \quad y_0 = e^{i\alpha}|y_0|, \quad \xi, \eta \in \mathbb{R}.$$

The solution is given by

$$y(t) = e^{(\eta + i\xi)t} e^{i\alpha}|y_0| = e^{i(\xi t + \alpha)} e^{\eta t}|y_0|;$$

thus,

$$|y(t)| = e^{\eta t}|y_0|, \quad y(t) = e^{i(\xi t + \eta)}|y(t)|.$$

Therefore, depending on the sign of η, the amplitude $|y(t)|$ of the solution grows, decays, or remains constant. The phase $\xi t + \eta$ is a linear function of t and changes rapidly if $|\xi|$ is large.

Next, consider the inhomogeneous problem

$$\frac{dy}{dt} = \lambda y + ae^{\mu t},$$
$$y(0) = y_0,$$

(1.5)

where λ, a, μ, and y_0 are complex constants. Regardless of the initial condition at first, we look for a *particular solution* of the form

$$y_P(t) = e^{\mu t}A, \quad A = \text{const.}$$

(1.6)

Introducing (1.6) into the differential equation (1.5) gives us

$$\mu e^{\mu t}A = \lambda e^{\mu t}A + ae^{\mu t},$$

that is,

$$(\mu - \lambda)A = a.$$

If $\mu \neq \lambda$ we obtain the particular solution

$$y_P(t) = e^{\mu t} \frac{a}{\mu - \lambda}.$$

On the other hand, if $\mu = \lambda$, the procedure above is not successful and we try to find a solution of the form[1]

$$y_P(t) = te^{\mu t} A. \tag{1.7}$$

Introducing (1.7) into the differential equation gives us

$$e^{\mu t} A + \mu te^{\mu t} A = \lambda te^{\mu t} A + ae^{\mu t}.$$

The last equation is satisfied if we choose $A = a$; recall that $\lambda = \mu$ by assumption. Let us summarize our results.

Lemma 1.1 *The function*

$$y_P(t) = \begin{cases} \dfrac{ae^{\mu t}}{\mu - \lambda} & \text{if } \mu \neq \lambda \\[2em] ate^{\mu t} & \text{if } \mu = \lambda \end{cases}$$

is a solution of the differential equation $dy/dt = \lambda y + ae^{\mu t}$.

Note that the particular solution $y_P(t)$ does not adjust, in general, to the initial data given (i.e., $y_P(0) \neq y_0$). The initial value problem (1.5) can now be solved in the following way. We introduce the dependent variable u by

$$y = y_P + u.$$

Initial value problem (1.5) yields

$$\frac{dy_P}{dt} + \frac{du}{dt} = \lambda(y_P + u) + ae^{\mu t},$$
$$y_P(0) + u(0) = y_0,$$

and, since $dy_P/dt = \lambda y_P + ae^{\mu t}$, we obtain

$$\frac{du}{dt} = \lambda u,$$
$$u(0) = y_0 - y_P(0).$$

Thus, $u(t)$ satisfies the corresponding homogeneous differential equation, and (1.3) yields

$$u(t) = e^{\lambda t}(y_0 - y_P(0)).$$

---

[1] The exceptional case, $\mu = \lambda$, is called the *case of resonance*.

The complete solution

$$y(t) = y_P(t) + u(t)$$

consists of two parts, $y_P(t)$ and $u(t)$. The function $y_P(t)$ is also called the *forced solution* because it has essentially the same behavior as that of the *forcing* $ae^{\mu t}$. The other part, $u(t)$, is often called the *transient solution* since it converges to zero for $t \to \infty$ if $\mathrm{Re}\,\lambda < 0$.

Finally, we want to show how we can solve the initial value problem

$$\frac{dy}{dt} = \lambda y + F(t),$$
$$y(0) = y_0, \tag{1.8}$$

with a general forcing $F(t)$. We can solve this problem by applying a procedure known as *Duhamel's principle*.

1.1.1   Duhamel's principle

Lemma 1.2 *The solution of* (1.8) *is given by*

$$y(t) = e^{\lambda t} y_0 + \int_0^t e^{\lambda(t-s)} F(s)\, ds. \tag{1.9}$$

Proof: Define $y(t)$ by formula (1.9). Clearly, $y(0) = y_0$ (i.e., the initial condition is satisfied). Also, $y(t)$ is a solution of the differential equation, because

$$\frac{dy}{dt} = \lambda e^{\lambda t} y_0 + F(t) + \lambda \int_0^t e^{\lambda(t-s)} F(s)\, ds = \lambda y(t) + F(t).$$

This proves the lemma.   ∎

Exercise 1.1   *Prove that the solution* (1.9) *is the unique solution of* (1.8).

We shall now discuss the relation between the solution to inhomogeneous equation (1.8) and the homogeneous equation

$$\frac{du}{dt} = \lambda u. \tag{1.10}$$

We consider (1.10) with initial condition $u = u(s)$ at a time $s > 0$. At a later time $t \geq s$ the solution is

$$u(t) = e^{\lambda(t-s)} u(s).$$

Thus, $e^{\lambda(t-s)}$ is a factor that connects $u(t)$ with $u(s)$. We will call it the *solution operator* and use the notation

$$S(t, s) = e^{\lambda(t-s)}, \quad \text{i.e.,} \quad u(t) = S(t, s) u(s). \tag{1.11}$$

The solution operator has the following properties:

$$S(t,0) = e^{\lambda t}, \quad S(t,t) = 1,$$
$$S(t,t_1)S(t_1,s) = S(t,s). \tag{1.12}$$

Now we can show that the solution of inhomogeneous equation (1.8) can be expressed in terms of the solution of homogeneous equation (1.10). Then (1.9) becomes

$$y(t) = S(t,0)y(0) + \int_0^t S(t,s)F(s)\,ds. \tag{1.13}$$

In a somewhat loose way, we may consider the integral as a "sum" of many terms $S(t,s_j)F(s_j)\Delta s$; think of approximating the integral by a Riemann sum. Then (1.13) expresses the solution of inhomogeneous problem (1.8) as a weighted superposition of solutions $t \to S(t,s)$ of homogeneous equation (1.10). The idea of expressing the solution of an inhomogeneous problem via solutions of the homogeneous equation is very useful. As we will see, it generalizes to systems of equations, to partial differential equations, and also to difference approximations. It is known as *Duhamel's principle*.

Exercise 1.2 *Use Duhamel's principle to derive a representation for the solution of*

$$\frac{dy}{dt} = \lambda(t)y + F(t),$$
$$y(0) = y_0.$$

Exercise 1.3 *Consider the inhomogeneous initial value problem*

$$\frac{dy}{dt} = \lambda y + P_n(t)\,e^{\mu t},$$
$$y(0) = y_0, \qquad \lambda, \mu, y_0 \in \mathbb{C},$$

where $P_n(t)$ is a polynomial of degree n with complex coefficients. Show that the solution to the problem is of the form

$$y(t) = e^{\lambda t}\tilde{y}_0 + Q_m(t)e^{\mu t},$$

where $Q_m(t)$ is a polynomial of degree m with $m = n$ in the nonresonance case ($\mu \neq \lambda$) and $m = n+1$ in the resonance case ($\mu = \lambda$). Determine $\tilde{y}_0$ in each case.

We now want to consider scalar equations with smooth variable coefficients, which leads to the next principle.

1.1.2 Principle of frozen coefficients

In many applications the problem with smooth variable coefficients can be *localized*, that is, it can be decomposed in many constant coefficient problems (by using a partition of unity). Then by solving all these constant coefficient problems, one can construct an approximate solution to the original variable coefficient problem. The approximation can be as good as one wants. The general theory concludes that if all relevant constant coefficient problems have a solution, the variable coefficient problem also has a solution. This procedure is known as the *principle of frozen coefficients.* We do not go into it more deeply here.

1.2   Variable coefficient linear equations

1.2.1   Principle of superposition

The initial value problem

$$\frac{dy}{dt} = a(t)y + F(t),$$
$$y(0) = y_0,$$

(1.14)

is an example of a linear problem. It has the following properties:

1. Let $y(t)$ be a solution of (1.14). Let σ be a constant and replace $F(t)$ and y_0 by $\sigma F(t)$ and σy_0, respectively. In other words, consider the new problem

$$\frac{d\tilde{y}}{dt} = a(t)\tilde{y} + \sigma F(t),$$
$$\tilde{y}(0) = \sigma y_0.$$

(1.15)

   Multiplying (1.14) by σ gives us

$$\frac{d(\sigma y)}{dt} = a(t)(\sigma y) + \sigma F(t),$$
$$(\sigma y(0)) = \sigma y_0.$$

   Thus, (1.15) solves the new problem and, using uniqueness, the solution (1.15) is

$$\tilde{y}(t) = \sigma y(t).$$

2. Consider (1.14) with a set of two forcing functions $F_1(t)$, $F_2(t)$ and two initial data y_{01}, y_{02}. Denote the resulting solutions by $y_1(t)$, $y_2(t)$, respectively:

$$\frac{dy_1}{dt} = a(t)y_1(t) + F_1(t),$$
$$y_1(0) = y_{01},$$
$$\frac{dy_2}{dt} = a(t)y_2(t) + F_2(t),$$
$$y_2(0) = y_{02}.$$

Adding the equations, we find that

$$\frac{d(y_1 + y_2)}{dt} = a(t)(y_1 + y_2) + (F_1(t) + F_2(t)),$$

$$y_1(0) + y_2(0) = y_{01} + y_{02}.$$

Thus, the sum

$$\tilde{y}(t) = y_1(t) + y_2(t)$$

is a solution of

$$\frac{d\tilde{y}}{dt} = a(t)\tilde{y} + (F_1(t) + F_2(t)),$$

$$\tilde{y}(0) = y_{01} + y_{02}.$$

To summarize, for a linear problem such as (1.14), we can use superposition of solutions to obtain new solutions. This property of linear systems, known as the *superposition principle,* can be used to compose solutions with complicated forcing functions out of solutions of simpler problems. Consider, for example,

$$\frac{dy}{dt} = -y + \sin^2(t), \quad y(0) = y_0. \tag{1.16}$$

Since

$$\sin^2(t) = \left(\frac{1}{2i}(e^{it} - e^{-it})\right)^2 = -\frac{1}{4}(e^{2it} - 2 + e^{-2it})$$

consists of three terms, we solve three problems:

$$\frac{dy_1}{dt} = -y_1 - \frac{1}{4}e^{2it},$$

$$\frac{dy_2}{dt} = -y_2 + \frac{1}{2},$$

$$\frac{dy_3}{dt} = -y_3 - \frac{1}{4}e^{-2it}.$$

We do not yet impose initial conditions, so that we can choose simple particular solutions. By Lemma 1.1, particular solutions of the equations above are

$$y_{1P} = -\frac{e^{2it}}{4(1 + 2i)}, \quad y_{2P} = \frac{1}{2}, \quad y_{3P} = -\frac{e^{-2it}}{4(1 - 2i)}.$$

Using the superposition principle, we find that

$$\tilde{y} = y_{1P} + y_{2P} + y_{3P} = \frac{1}{2} - \frac{1}{4}\left(\frac{e^{2it}}{1 + 2i} + \frac{e^{-2it}}{1 - 2i}\right)$$

$$= \frac{1}{2} - \frac{1}{20}\left((1 - 2i)e^{2it} + (1 + 2i)e^{-2it}\right) = \frac{1}{2} - \frac{1}{10}\cos(2t) - \frac{1}{5}\sin(2t)$$

solves

$$\frac{d\tilde{y}}{dt} = -\tilde{y} + \sin^2(t).$$

Clearly,

$$\tilde{y}(0) = \tfrac{1}{2} - \tfrac{1}{10} = \tfrac{2}{5}.$$

Therefore, the solution of (1.16) is given by

$$y(t) = \tilde{y}(t) + \sigma e^{-t},$$

where σ is determined by the initial condition

$$y(0) = \tilde{y}(0) + \sigma = \tfrac{2}{5} + \sigma = y_0, \quad \text{i.e.,} \quad \sigma = y_0 - \tfrac{2}{5}.$$

The superposition principle relies only on linearity; it holds for any linear equation or system of linear equations, both ordinary and partial differential equations. An equation is linear if the dependent variable and its derivatives appear linearly only (i.e., as the first power), in the equation.

Exercise 1.4 *Solve the initial value problem*

$$\frac{dy}{dt} = -2y + (1 + t^2)\sin(t)\cos(t),$$
$$y(0) = 1.$$

1.2.2 Duhamel's principle for variable coefficients

We want to discuss now the solution of problem (1.14) in terms of Duhamel's principle. To this end we discuss the solution operator in a more abstract setting.

Consider first an initial value problem for the homogeneous equation associated with (1.14):

$$\frac{dy}{dt} = a(t)y, \tag{1.17}$$
$$y(0) = y_0.$$

The *solution operator* for problem (1.17) is given by

$$S(t_2, t_1) = \exp\left(\int_{t_1}^{t_2} a(s)\, ds\right). \tag{1.18}$$

Clearly,

$$y(t) = S(t, 0)y_0$$

is the solution of (1.17). With this solution operator, Duhamel's principle [see equation (1.13)] generalizes to our variable coefficient problem (1.14):

$$y(t) = S(t, 0)y(0) + \int_0^t S(t, r)F(r)\, dr. \tag{1.19}$$

This can be proved in terms of general properties of the solution operator.
It is not difficult to show that (1.18) has the following properties:

1. $S(t,t) = I$. Here I represents the identity operator [i.e., $Iv(t) = v(t)$].

2. Let $t \geq t_1 \geq 0$. Then

$$S(t, t_1)S(t_1, 0) = S(t, 0).$$

3. $S(t, r)$ is a smooth function of t and

$$\frac{\partial S(t, r)}{\partial t} = a(t)S(t, r).$$

We shall use these properties to prove that (1.19) solves (1.14). Since $S(0,0) = I$,
we have $y(0) = y_0$. Also,

$$
\begin{aligned}
\frac{dy}{dt} &= S_t(t, 0)y(0) + F(t) + \int_0^t S_t(t, r)F(r)\,dr \\
&= a(t)\left(S(t, 0)y(0) + \int_0^t S(t, r)F(r)dr\right) + F(t) \\
&= a(t)y(t) + F(t).
\end{aligned}
$$

Therefore, $y(t)$ given by (1.19) is the solution of (1.14).

Exercise 1.5 *Find the solution operator and, using Duhamel's principle, the solution of the following initial value problems.*
    (a)

$$
\begin{aligned}
\frac{dy}{dt} &= \frac{y}{1+t} + t, \\
y(0) &= 1.
\end{aligned}
$$

(b)

$$
\begin{aligned}
\frac{dy}{dt} &= -2ty + 2te^{-2t^2}, \\
y(0) &= 0.
\end{aligned}
$$

(c)

$$
\begin{aligned}
\frac{dy}{dt} &= \cos(t)y + \sin(t)\cos(t), \\
y(0) &= 2.
\end{aligned}
$$

1.3   Perturbations and the concept of stability

Given a problem and perturbations to it, we want to know what effect the perturbations have on the solution.

As an example, consider the initial value problem

$$\frac{dy}{dt} = \lambda y - e^{-t},$$
$$y(0) = \frac{1}{\lambda + 1} \tag{1.20}$$

with $\lambda \neq -1$. (The exceptional case of resonance, $\lambda = -1$, can be treated with slight modifications.)

The solution of (1.20) is the decaying function

$$y(t) = \frac{e^{-t}}{\lambda + 1}.$$

Now consider the same differential equation with perturbed initial data

$$\frac{d\tilde{y}}{dt} = \lambda \tilde{y} - e^{-t},$$
$$\tilde{y}(0) = \frac{1}{\lambda + 1} + \varepsilon, \tag{1.21}$$

where $0 < \varepsilon \ll 1$ is a small constant. Let $w(t) = \tilde{y}(t) - y(t)$ denote the difference between the perturbed and original solutions. Subtracting (1.20) from (1.21), we obtain

$$\frac{dw}{dt} = \lambda w,$$
$$w(0) = \tilde{y}(0) - y(0) = \varepsilon,$$

whose solution is

$$w(t) = e^{\lambda t}\varepsilon, \quad \text{i.e.,} \quad |w(t)| = e^{\text{Re } \lambda t}|\varepsilon|. \tag{1.22}$$

Depending on the sign of $\text{Re } \lambda$, there are three possibilities.

1. $\text{Re } \lambda < 0$. In this case the perturbation term $w(t)$ decays exponentially with time and the solution of the perturbed problem converges to the solution of the original problem as t increases. In Figure 1.5 the solid line represents the solution with $\lambda = -7/12$, and the dashed line is the solution with perturbed initial data.

2. $\text{Re } \lambda = 0$. The perturbation $w(t)$ does not decrease with time, but it does not grow either (see Figure 1.6).

3. $\text{Re } \lambda > 0$. The perturbation grows exponentially in time. Figure 1.7 shows the perturbed and unperturbed solutions for $\lambda = 1$.

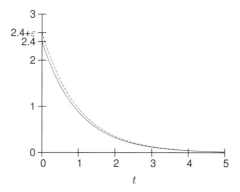

Figure 1.5   Decaying perturbation.

In the latter case it will be very difficult to compute the original solution accurately in long time intervals. For example, if $\lambda = 1$ and $\varepsilon = 10^{-10}$, then

$$w(T) = e^T 10^{-10} = 1 \quad \text{if} \quad T = \log_e 10^{10} = 10 \log_e 10 \simeq 25.$$

Therefore, if the calculation introduces an error $\varepsilon = 10^{-10}$ at $t = 0$, this error will grow to $w(T) = 1$ at about $T = 25$. This growth holds even if no further errors, except the original error $\varepsilon = 10^{-10}$ at $t = 0$, are introduced.

In applications the initial data and the forcing are never given exactly. Therefore, if $\mathrm{Re}\,\lambda > 0$, one cannot guarantee that the answer computed is close to the correct answer. For $\mathrm{Re}\,\lambda < 0$, the situation is the opposite: Initial errors in the data are wiped out. Problems corresponding to $\mathrm{Re}\,\lambda < 0$ are called *strongly stable*. If $\mathrm{Re}\,\lambda = 0$, the problem is *stable* but not strongly stable, and if $\mathrm{Re}\,\lambda > 0$, the problem is *unstable*.

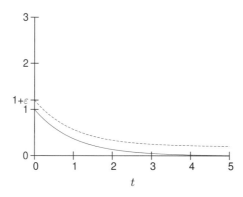

Figure 1.6   Non-decaying perturbation.

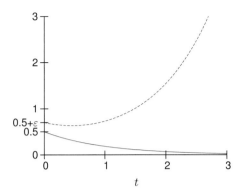

Figure 1.7   Exponentially growing perturbation.

Next, let us perturb the forcing and consider

$$\frac{d\tilde{y}}{dt} = \lambda\tilde{y} - e^{-t} + \varepsilon G(t),$$

$$\tilde{y}(0) = \frac{1}{\lambda + 1}$$

(1.23)

instead of (1.20). The error term $w(t) = \tilde{y}(t) - y(t)$ solves

$$\frac{dw}{dt} = \lambda w + \varepsilon G(t),$$

$$w(0) = 0,$$

(1.24)

and we obtain by Duhamel's principle,

$$w(t) = \varepsilon \int_0^t e^{\lambda(t-s)} G(s) \, ds.$$

Therefore, $w(t)$ satisfies the estimate

$$|w(t)| \leq \varepsilon \int_0^t |e^{\lambda(t-s)}||G(s)|ds \leq \varepsilon \max_{0 \leq s \leq t} |G(s)| \int_0^t e^{\text{Re}\,\lambda(t-s)} ds$$

$$= \varepsilon \max_{0 \leq s \leq t} |G(s)| \begin{cases} \dfrac{e^{\text{Re}\,\lambda t} - 1}{\text{Re}\,\lambda} & \text{if Re}\,\lambda \neq 0 \\[3mm] t & \text{if Re}\,\lambda = 0 \end{cases}$$

(1.25)

We arrive at essentially the same conclusions as those for the perturbed initial data:

1. If the problem is strongly stable (i.e., $\text{Re}\,\lambda < 0$), the perturbation of the solution is bounded by

$$|w(t)| \leq \frac{\varepsilon}{|\text{Re}\,\lambda|} \max_{0 \leq s \leq t} |G(s)|;$$

that is, the difference of the solutions is of the same order as the perturbation of the forcing.

2. If the problem is stable but not strongly stable (i.e., $\operatorname{Re} \lambda = 0$), we obtain

$$|w(t)| \leq \varepsilon \int_0^t |G(s)|\, ds \leq \varepsilon t \max_{0 \leq s \leq t} |G(t)|.$$

Thus, the difference in the solutions can be estimated in terms of the integrated effect of the perturbation. This effect typically grows linearly with time. In most applications one can handle such situations and obtain accurate solutions by keeping the perturbations sufficiently small. For example, if $\varepsilon = 10^{-10}$ and $|G(t)| \leq 1$, it takes a very long time before the effect of the perturbation is noticed.

3. If $\operatorname{Re} \lambda > 0$, the effect of the perturbation grows exponentially in time. In a long time interval, this may change the true solution drastically.

There are no difficulties in generalizing this observation to linear equations with variable coefficients:

$$\frac{dy}{dt} = \lambda(t)y + F(t), \tag{1.26}$$
$$y(0) = y_0.$$

The influence of perturbations of the forcing and of the initial data depends on the behavior of the solution operator

$$S(t, s) = e^{\int_s^t \lambda(\eta)d\eta}, \quad t \geq s.$$

Definition 1.3 *Consider the linear initial value problem* (1.26) *and its solution operator* $S(t, s)$. *The problem is called strongly stable, stable, or unstable if the solution operator satisfies, respectively, the following estimates:*

$$|S(t, s)| \leq e^{-\delta(t-s)}, \quad |S(t, s)| \leq \text{const.}, \quad or \quad |S(t, s)| \geq e^{\delta(t-s)},$$

where δ is a positive constant.

Exercise 1.6 *Consider, instead of* (1.20), *the initial value problem for the resonance case*

$$\frac{dy}{dt} = \lambda y + e^{\lambda t},$$
$$y(0) = 0,$$

and the problem with perturbed initial data,

$$\frac{d\tilde{y}}{dt} = \lambda \tilde{y} + e^{\lambda t},$$
$$\tilde{y}(0) = \varepsilon, \quad 0 < \varepsilon \ll 1.$$

Show that the same conclusions of the nonresonance case can be drawn for $w(t) = \tilde{y} - y(t)$.

1.4   Nonlinear equations: the possibility of blow-up

Nonlinearities in the equation can produce a solution that blows up in finite time, that is, a solution that does not exist for all times. Consider, for example, the nonlinear initial value problem given by

$$\frac{dy}{dt} = y^2, \quad t \geq 0,$$
$$y(0) = y_0.$$

(1.27)

For $y_0 = 0$ the solution is $y = 0$ for all times. Therefore, assume that $y_0 \neq 0$ in the following. To calculate the solution, we write the differential equation in the form

$$\frac{1}{y^2}\frac{dy}{dt} = 1$$

and integrate:

$$\int_0^t \frac{1}{y^2(s)}\frac{dy}{ds}\, ds = t.$$

The change of variables $y(s) = v$ gives us

$$\int_{y_0}^{y(t)} \frac{dv}{v^2} = t,$$

and we obtain

$$\frac{1}{y_0} - \frac{1}{y(t)} = t, \quad \text{i.e.,} \quad y(t) = \frac{1}{(1/y_0) - t}.$$

For $y_0 > 0$ the solution blows up at $t = 1/y_0$ (see Figure 1.8). This blow-up or divergence of the solution at a finite time is a consequence of the nonlinearity, that is, the term y^2 on the right-hand side of the equation. This behavior cannot occur in

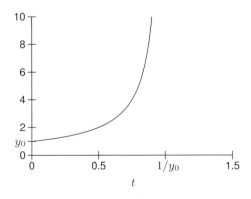

Figure 1.8 $y_0 > 0$.

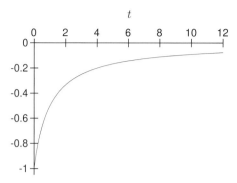

Figure 1.9 $y_0 < 0$.

a linear problem. On the other hand, if $y_0 < 0$, the solution $y(t)$ exists for all $t \geq 0$ and converges to zero for $t \to \infty$ (see Figure 1.9).

Consider now the more general problem

$$\frac{dy}{dt} = f(y, t),$$

$$y(t_0) = y_0.$$

(1.28)

We give here without proof a simple version of the classical *existence and uniqueness theorem* for scalar ordinary differential equations (see, e.g., [3], chapt. 5).

Theorem 1.4 *If $f(y, t)$ and $\partial f(y, t)/\partial y$ are continuous functions in a rectangle $\Omega = [y_0 - b, y_0 + b] \times [t_0 - a, t_0 + a]$, $a, b > 0$, and $|f(y, t)| \leq M$ on Ω, there exists a unique, continuously differentiable solution $y(t)$ to the problem (1.28) in the interval $|t - t_0| \leq \Delta t = \min\{a, b/M\}$.*

Remark 1.5 *The time interval of existence depends on how large one can choose the rectangle, and so on the initial point (y_0, t_0). The solution can be continued to the future by solving the equation with new initial conditions starting at the point $(y(t_0 + \Delta t), t_0 + \Delta t)$. If one tries to continue the solution as much as possible, there are two possibilities:*

1. *One can continue the solution to arbitrarily large times, that is, the solution exists for all times $t \geq 0$.*

2. *There is a finite time $T_0 = T_0(y_0) > 0$ such that the solution exists for all times $t < T_0$ but not at T_0. In this case the solution blows up at T_0; that is, $\lim_{t \to T_0^-} |y(t)| = \infty$.*

Exercise 1.7 *Show that if a smooth solution $y(t)$ to (1.28) blows up at finite time, its derivative dy/dt blows up at the same time. Hint: Use the mean value theorem.*

Exercise 1.8 *Show that the converse of exercise (1.7) is false. To this end, consider the initial value problem*

$$\frac{dy}{dt} = \frac{1}{2-y},$$
$$y(0) = 1.$$

Explicitly find the solution $y(t)$ and check that $dy/dt \to \infty$ when $t \to (\frac{1}{2})^-$ and that, nevertheless, $y(t)$ stays bounded.

Exercise 1.9 *Is it possible that the solution of the real equation*

$$\frac{dy}{dt} = \sin(y),$$
$$y(0) = y_0$$

blows up at a finite time? Explain the Answer.

1.5   Principle of linearization

Consider the initial value problem

$$\frac{dy}{dt} = \lambda y + y^2 + F(t),$$
$$y(0) = 0.$$
(1.29)

Assume that

$$F(t) = \cos(t) - \lambda \sin(t) - \sin^2(t).$$

A simple calculation shows that the solution of (1.29) is given by

$$y(t) = \sin(t).$$

Let ε with $0 \le \varepsilon \ll 1$ be a small constant and consider the perturbed problem

$$\frac{d\tilde{y}}{dt} = \lambda \tilde{y} + \tilde{y}^2 + F(t) + \varepsilon G(t),$$
$$\tilde{y}(0) = 0.$$
(1.30)

Here $G(t)$ is a smooth function with

$$|G(t)| + \left| \frac{dG}{dt} \right| \le 1.$$
(1.31)

By Section 1.3 we expect that in some time interval $0 \le t \le T$,

$$\tilde{y}(t) = \sin(t) + \mathcal{O}(\varepsilon).^2$$

---

[2]From now on we frequently use the notation $\mathcal{O}$; for a precise definition, consult Section A.2.

Therefore, we make the following change of variables:

$$\tilde{y}(t) = \sin(t) + \varepsilon u(t). \tag{1.32}$$

Introducing (1.32) into (1.30) gives us

$$\cos(t) + \varepsilon \frac{du}{dt} = \lambda \sin(t) + \varepsilon \lambda u + \sin^2(t) + 2\varepsilon \sin(t)u + \varepsilon^2 u^2 + F(t) + \varepsilon G(t).$$

The form of F gives us

$$\frac{du}{dt} = (\lambda + 2\sin t)u + \varepsilon u^2 + G(t),$$
$$u(0) = 0. \tag{1.33}$$

We expect that $|u| \leq 1$ in some time interval $0 \leq t \leq T$, and therefore we can neglect the quadratic term εu^2 to obtain the *linearized equation*

$$\frac{d\tilde{u}}{dt} = (\lambda + 2\sin t)\tilde{u} + G(t),$$
$$\tilde{u}(0) = 0. \tag{1.34}$$

In Section 1.3 we discussed the growth behavior of the solutions of (1.34). It depends on the solution operator

$$S(t, s) = e^{\lambda(t-s) + 2 \int_s^t \sin(r)\, dr}.$$

Exercise 1.10 *Prove, using the solution operator $S(t, s)$, that the problem is strongly stable for $\lambda < 0$.*

If the linearized equation is strongly stable (i.e., $\mathrm{Re}\,\lambda < 0$), then

$$|\tilde{u}(t)| \leq \text{const.} \max_{0 \leq s \leq t} |G(s)|$$

and, using (1.31), $\tilde{u}$ is bounded for all times. In this case one can also show that, for sufficiently small ε,

$$|u - \tilde{u}| = \mathcal{O}(\varepsilon)$$

for all times. Thus, the linearized equation determines, to first approximation, the effect of the perturbation on the solution.

If the linearized equation is only stable, then

$$|\tilde{u}(t)| \leq \text{const.} \int_0^t |G(s)|\, ds$$

and

$$|u - \tilde{u}| \leq \text{const.}\,\varepsilon \int_0^t |G(s)|\, ds,$$

provided that

$$\varepsilon \max_{0 \le s \le t} |\tilde{u}(s)|^2 \ll 1.$$

Therefore, the linearized equation describes the behavior of the perturbation in every time interval $0 \le t \le T$ with $T^2 \varepsilon \ll 1$.

If the linearized equation is unstable,

$$|\tilde{u}(t)| \le \text{const.} \, e^{\text{Re} \, \lambda t}$$

and the time interval where the linearized equation (1.34) is a good approximation of (1.33) is restricted to a time interval $0 \le t \le T$ with

$$T = \mathcal{O}(\log_e \varepsilon^{-1}).$$

This behavior is general in nature.

Consider the nonlinear equation

$$\frac{dy}{dt} = f(y, t), \tag{1.35}$$
$$y(0) = y_0.$$

Assume that the solution $y(t)$ of this problem is known. Consider a perturbation

$$\frac{d\tilde{y}}{dt} = f(\tilde{y}, t) + \varepsilon G(t), \tag{1.36}$$
$$\tilde{y}(0) = y_0 + \varepsilon \tilde{y}_0.$$

We make the change of variables

$$\tilde{y} = y + \varepsilon u.$$

Since

$$f(y + \varepsilon u, t) = f(y, t) + \varepsilon \frac{\partial f}{\partial y}(y, t)u + \varepsilon^2 \mathcal{O}(u^2),$$

we obtain

$$\frac{dy}{dt} + \varepsilon \frac{du}{dt} = f(y, t) + \varepsilon \frac{\partial f}{\partial y}(y, t)u + \varepsilon^2 \mathcal{O}(u^2) + \varepsilon G(t), \tag{1.37}$$
$$u(0) = \tilde{y}_0.$$

Neglecting the quadratic terms, we obtain the linearized equation

$$\frac{du}{dt} = \frac{\partial f}{\partial y}(y, t)u + G(t), \tag{1.38}$$
$$u(0) = \tilde{y}_0.$$

The effect of the perturbation depends on the stability properties of (1.38).

Linearization is a very important tool because it is used to show that the nonlinear problem has a unique solution locally.

CHAPTER 2

METHOD OF EULER

Essentially all concepts introduced in Chapter 1 have their counterpart in this chapter. We introduce the simplest difference method that one can use to approximate an initial value problem. Most of the fundamental concepts needed to understand numerical approximations to differential equations follow the concepts introduced here for the method of Euler. These concepts are more complicated; therefore, one should understand the procedures for differential equations before one starts with the difference approximations. New concepts are accuracy, truncation error, and stability region. Also, a discrete version of Duhamel's principle is discussed. General one-step methods are defined and a section is devoted to tests of correctness of a program.

2.1 Explicit Euler method

There are many methods of computing approximations to the solution of an initial value problem. The explicit Euler method,[1] which we discuss here, is the most easily

---

[1] Also called the Euler forward method, or simply the Euler method.

Introduction to Numerical Methods for Time Dependent Differential Equations, First Edition.
By Heinz-O. Kreiss and Omar E. Ortiz. Copyright © 2014 John Wiley & Sons, Inc.

Figure 2.1   Grid starting at $t = 0$.

understood and implemented. To describe it, let us first introduce the concepts of grid and grid function.

Starting form the initial time in our problem, say $t = 0$, we divide the t-axis into subintervals of length $k > 0$ and obtain a *grid* (see Figure 2.1). The endpoints $t_n = nk, n = 0, 1, 2, \ldots$ of the subintervals are called *grid points*. A *grid function* $g_n = g(nk)$ is a function that is defined on the grid. The explicit Euler method for an initial value problem

$$\frac{dy}{dt} = f(y, t),$$
$$y(0) = y_0,$$

will be derived below, but we start with the particular example

$$\frac{dy}{dt} = \lambda y + ae^{\mu t},$$
$$y(0) = y_0, \tag{2.1}$$

and use it for an error discussion.

For simplicity, let us assume that $\mu \neq \lambda$ (i.e., we do not have a case of resonance). By Section 1.1 we know the exact solution,

$$y = y_P + u, \quad \text{with} \quad y_P = \frac{ae^{\mu t}}{\mu - \lambda}, \quad u = e^{\lambda t}u(0), \quad u(0) = y_0 - y_P(0). \tag{2.2}$$

Let us now replace (2.1) by a difference approximation. By Taylor expansion, we have for any smooth function $y(t)$,

$$y_{n+1} = y_n + k\left(\frac{dy}{dt}\right)_n + \mathcal{O}(k^2). \tag{2.3}$$

If $y(t)$ is the solution of the differential equation (2.1), we can replace dy/dt by $\lambda y + e^{\mu t}$ and obtain

$$y_{n+1} = y_n + k(\lambda y_n + ae^{\mu t_n}) + \mathcal{O}(k^2).$$

If we neglect the $\mathcal{O}(k^2)$ term and write v_n instead of y_n, we obtain the *explicit Euler method* for (2.1):

$$v_{n+1} = v_n + k(\lambda v_n + ae^{\mu t_n}) = (1 + k\lambda)v_n + kae^{\mu t_n}, \quad n = 0, 1, 2, \ldots, \tag{2.4}$$
$$v_0 = y_0. \tag{2.5}$$

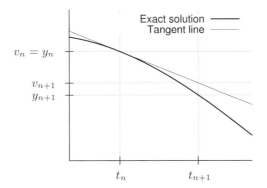

Figure 2.2   Explicit Euler method starting with $v_n = y_n$.

It is clear that the difference equation (2.4) can be solved recursively: Once v_n is known for some n, (2.4) can be used to calculate v_{n+1}, etc. Thus, (2.4) and (2.5) determine a unique grid function v_n, $n = 0, 1, 2, \ldots$. It is also clear that this grid function depends on the *step size* k, but this is suppressed in our notation.

The exact solution to an initial value problem can be thought, geometrically, to be the integral curve, in the (t, y) plane, of the direction field defined by the equation that passes through the initial point (t_0, y_0). One can think of Euler's explicit method, geometrically, as a method that computes the point (t_{n+1}, v_{n+1}) by following the tangent line to the graph of the exact solution that passes though the point (t_n, v_n) (see Figure 2.2).

It is not difficult to derive the explicit Euler method for general equations:

$$\frac{dy}{dt} = f(y, t),$$
$$y(0) = y_0.$$
(2.6)

Taylor expansion gives us

$$y(t_{n+1}) = y(t_n) + k\frac{dy}{dt}(t_n) + \frac{k^2}{2}\frac{d^2y}{dt^2}(t_n) + \mathcal{O}(k^3)$$
$$= y(t_n) + kf(y(t_n), t_n) + \frac{k^2}{2}\frac{dy}{dt}(t_n) + \mathcal{O}(k^3).$$

Neglecting terms of order $\mathcal{O}(k^2)$, we obtain the explicit Euler method.

Definition 2.1 *The explicit Euler method (or simply the Euler method) that approximates problem (2.6) is*

$$v_{n+1} = v_n + kf(v_n, t_n), \quad t_n = nk.$$
$$v_0 = y_0,$$

2.2 Stability of the explicit Euler method

Consider the initial value problem

$$\frac{dy}{dt} = \lambda y + F(t),$$
$$y(0) = y_0.$$

(2.7)

If one approximates this problem with Euler's method, one expects that the difference between the approximate solution and the exact solution to (2.7) will approach zero as $k \to 0$. However, in actual numerical computations, one might choose one or a few step sizes but cannot send $k \to 0$. In a concrete application, how should one choose k? There are two aspects that one needs to consider to make a good choice: the *stability* and the *accuracy* of the approximation. The latter aspect is determined by the size of the *truncation error.*

We start the discussion of these concepts with an example. Consider the initial value problem

$$\frac{dy}{dt} = -100y,$$
$$y(0) = 1.$$

(2.8)

Let us approximate (2.8) by the explicit Euler method:

$$v_{n+1} = (1 - 100k)v_n,$$
$$v_0 = 1.$$

(2.9)

The exact solution of (2.8),

$$y(t) = e^{-100t},$$

decays rapidly to zero for increasing t. On the other hand, if $k > 2/100$, the numerical solution v_n increases in absolute value with increasing n and does not resemble y_n at all. For the example above, a reasonable requirement is to choose k so small the the solution v_n of the difference approximation does not grow in absolute value with increasing n. Thus, one needs to choose $k \leq 2/100$.

For the more general case,

$$\frac{dy}{dt} = \lambda y,$$
$$y(0) = 1, \quad \lambda \in \mathbb{C}, \operatorname{Re}\lambda \leq 0,$$

(2.10)

we have the following result.

Lemma 2.2 *Approximate (2.10) by the explicit Euler method:*

$$v_{n+1} = (1 + \lambda k)v_n,$$
$$v_0 = y_0.$$

If $y_0 \neq 0$, we have

$$|v_n| \leq |v_0| \quad for \quad n = 1, 2, 3, \ldots \tag{2.11}$$

if and only if

$$|1 + \lambda k| \leq 1. \tag{2.12}$$

Proof: Clearly, the solution v_n of the difference equation is given by

$$v_n = (1 + \lambda k)^n y_0.$$

If $|1 + \lambda k| > 1$, the absolute value of v_n grows with increasing n. On the other hand, if $|1 + \lambda k| \leq 1$, then (2.11) holds. ∎

Definition 2.3 *The set of all complex numbers $\mu = \lambda k$ for which the estimate* (2.12) *holds is called the* stability region *of the explicit Euler method.*

Clearly, for the explicit Euler method, the stability region consists of a closed disk with radius $\lambda k = 1$ and center in $\lambda k = -1$ in the complex $\mu = \lambda k$ plane (see Figure 2.3).

2.3 Accuracy and truncation error

We start our discussion about the accuracy of the Euler approximation by defining the truncation error.

Definition 2.4 *Consider the initial value problem*

$$\frac{dy}{dt} = f(y, t),$$

$$y(0) = y_0,$$

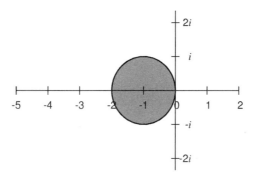

Figure 2.3 The shadowed disk is the stability region of the explicit Euler method in the complex plane.

and approximate it by the explicit Euler method:

$$v_{n+1} = v_n + kf(v_n, t_n), \quad v_0 = y_0.$$

Let $y(t)$ denote the solution of the initial value problem in some interval $0 \le t \le T$ and substitute $y_n = y(t_n)$ into the difference equations. Then the truncation error R_n for this method is defined as

$$R_n = \frac{y_{n+1} - y_n}{k} - f(y_n, t_n). \tag{2.13}$$

Remark 2.5 *It is clear that the truncation error R_n depends on the step size k and on the solution $y(t)$ under consideration. We usually suppress this in our notation.*

The truncation error measures how well the solution $y(t)$ of the differential equation satisfies the difference equations that are used to compute the numerical approximation v_n. kR_n represents the terms that we have neglected when deriving the method.

As an example, consider the initial value problem

$$\frac{dy}{dt} = \lambda y + F(t),$$
$$y(0) = y_0, \tag{2.14}$$

for a linear equation, and approximate it by the explicit Euler method:

$$v_{n+1} = (1 + \lambda k)v_n + kF_n,$$
$$v_0 = y_0. \tag{2.15}$$

By Taylor expansion,

$$y(t_n + k) = y(t_n) + k\frac{dy}{dt}(t_n) + \frac{k^2}{2}\frac{d^2y}{dt^2}(t_n) + \mathcal{O}(k^3). \tag{2.16}$$

Therefore,

$$y_{n+1} - (1 + \lambda k)y_n - kF_n = kR_n \tag{2.17}$$

with

$$kR_n = y_n + k\left(\frac{dy}{dt}\right)_n + \frac{k^2}{2}\left(\frac{d^2y}{dt^2}\right)_n - (1 + \lambda k)y_n - kF_n + \mathcal{O}(k^3)$$
$$= \frac{k^2}{2}\left(\frac{d^2y}{dt^2}\right)_n + \mathcal{O}(k^3). \tag{2.18}$$

Thus, the method is first-order accurate, and for $k \to 0$ the truncation error converges to zero. The convergence is linear with k and uniform in t; that is, as the analytic solution $y(t)$ is smooth in a given time interval $t \in [0, T]$ (in this linear example, T can be chosen as large as one pleases), we have

$$|R_n| = \left|\frac{1}{2}\frac{d^2y}{dt^2}(t_n) + \mathcal{O}(k)\right| k \le C(T)k.$$

Thus,

$$\max_{0 \le t_n \le T} |R_n| = \mathcal{O}(k),$$

and we say that the explicit Euler method is accurate of order 1.

2.4 Discrete Duhamel's principle and global error

We now relate the stability and truncation error to the global (actual) error $e_n = y_n - v_n$. We start by considering the linear initial value problem of the example given in Section 2.3. Subtracting (2.15) from (2.17), we obtain

$$e_{n+1} - (1 + \lambda k)e_n = kR_n.$$

Thus, the error is determined recursively by

$$e_{n+1} = (1 + \lambda k)e_n + kR_n,$$
$$e_0 = 0. \tag{2.19}$$

From this recursive error equation we can derive an explicit formula for the error. The result (2.21), given below, is analogous to (1.13) and constitutes an example of Duhamel's principle for difference equations.

Let

$$S_k(n, m) = (1 + \lambda k)^{n-m}, \quad n \geq m, \tag{2.20}$$

denote the solution operator of the *homogeneous* difference equations corresponding to (2.19). We have

Lemma 2.6 *The solution e_n of the error equation (2.19 has, in terms of the solution operator (2.20), the representation*

$$e_n = S_k(n, 0)e_0 + k \sum_{m=0}^{n-1} S_k(n - 1, m)R_m, \quad n \geq 1. \tag{2.21}$$

Proof: We use induction in n. For $n = 1$ the formula is true, because

$$e_1 = (1 + \lambda k)e_0 + kR_0.$$

Assuming that (2.21) holds for n, we now show its validity for $n + 1$. First note that

$$(1 + \lambda k)S_k(n - 1, m) = S_k(n, m)$$

and

$$S(n, n) = 1.$$

By (2.19) and (2.21), we have

$$e_{n+1} = (1 + \lambda k)e_n + kR_n$$

$$= (1 + \lambda k)\left(S_k(n, 0)e_0 + k \sum_{m=0}^{n-1} S(n - 1, m)R_m\right) + kR_n$$

$$= S_k(n + 1, 0)e_0 + k \sum_{m=0}^{n-1} S_k(n, m)R_m + kS(n, n)R_n$$

$$= S_k(n + 1, 0)e_0 + k \sum_{m=0}^{n} S_k(n, m)R_m,$$

which is the correct expression for e_{n+1}.   ∎

Let us now assume that $\mathrm{Re}\,\lambda < 0$ in example (2.14). Then, if we choose the step size k small enough, we have $|1 + \lambda k| \leq 1$ (i.e., λk lies in the stability region of the method). Under this assumption we can estimate the error.

Theorem 2.7 *Assume that λk belongs to the stability region of the method. Then the error $e_n = y_n - v_n$ satisfies the estimate*

$$|e_n| \leq k \sum_{m=0}^{n-1} |R_m| = \frac{k^2}{2} \sum_{m=0}^{n-1} \left|\left(\frac{d^2y}{dt^2}\right)_m\right| + \mathcal{O}(k^3). \tag{2.22}$$

Proof: First note that $e_0 = 0$ in (2.21). If λk belongs to the stability region, $|S_k(n-1, m)| \leq 1$, and the estimate follows from (2.21) and (2.18).   ∎

Remark 2.8 On local and global accuracy. *Since a term of order $\mathcal{O}(k^2)$ has been neglected in deriving the Euler method, one says that the method is "locally accurate of order k^2", that is, if the exact solution to the equation and its Euler approximation coincide at time t, the difference between them at time $t + k$ vanishes quadratically with k as k goes to zero. On the other hand, the global error in any finite interval is only $\mathcal{O}(k)$ and not $\mathcal{O}(k^2)$. This is so since to get to time $t \simeq 1$, the right-hand side of (2.22) adds $\mathcal{O}(1/k)$ terms of order $\mathcal{O}(k^2)$ and $\mathcal{O}(k^2)\mathcal{O}(1/k) = \mathcal{O}(k)$.*

The estimate for the global error just obtained is not necessarily sharp: the best possible estimate that one can get for a linear problem. However, it has the property that one can generalize it to nonlinear problems. To see this, let us estimate the error when one approximates a nonlinear problem:

$$\begin{aligned}\frac{dy}{dt} &= f(y, t), \\ y(0) &= y_0,\end{aligned} \tag{2.23}$$

where the function $f(y, t)$ is assumed to be smooth. Assume that the solution $y(t)$ of (2.23) exists and is smooth in some time interval $[0, T]$. Let us approximate our problem with the explicit Euler method

$$\begin{aligned}v_{n+1} &= v_n + kf(v_n, t_n), \\ v(0) &= y_0.\end{aligned}$$

As above, we have for the truncation error,

$$kR_n = y_{n+1} - y_n - kf(y_n, t_n) = \frac{k^2}{2}\left(\frac{d^2y}{dt^2}\right)_n + \mathcal{O}(k^3).$$

Thus, the error $e_n = y_n - v_n$ solves

$$\begin{aligned}e_{n+1} &= e_n + k\Big(f(y_n, t_n) - f(v_n, t_n)\Big) + kR_n, \\ e_0 &= 0.\end{aligned}$$

By Taylor expansion,

$$f(y_n, t_n) = f(v_n + e_n, t_n)$$
$$= f(v_n, t_n) + \frac{\partial f}{\partial y}(v_n, t_n)e_n + \mathcal{O}(e_n^2).$$

Neglecting the quadratic terms and using the notation $f_y(v_n, t_n) = \lambda_n$, we obtain

$$e_{n+1} = (1 + \lambda_n k)e_n + kR_n,$$
$$e_0 = 0.$$
(2.24)

The solution operator of the homogeneous equation corresponding to (2.24) is now

$$S_k(n, m) = \prod_{j=m}^{n-1} (1 + \lambda_j k).$$

With this definition of $S_k(n, m)$ the representation (2.21) of Lemma 2.6 holds. Assume now that $\operatorname{Re} \lambda_j < 0$ [which is an assumption on $df(y, t)/dy$] and choose k so that $\lambda_j k$ belongs to the stability region for all j. Then

$$|S_k(n, m)| \leq 1$$

and the error estimate of Theorem 2.7 also holds for the nonlinear case.

Exercise 2.1 *Consider the initial value problem*

$$\frac{dy}{dt} = \lambda y + F(t), \qquad 0 \leq t \leq 2,$$
$$y(0) = 1,$$

with $\lambda = -1$, $F(t) = \sin(2\pi t)$.

(a) Solve the problem analytically and discuss the behavior of the solution. Display the solution as a graph.

(b) Consider the explicit Euler method applied to this problem:

$$v_{n+1} = (1 + \lambda k)v_n + kF(t),$$
$$v_0 = 1$$

and write a computer program that implements this method for using three different time-step values: $k = 0.1, 0.01, 0.001$. *For each value of* k, *display the solution as a graph.*

(c) For each value of k, *compute the error* $\epsilon_n = |v_n - y(t_n)|$ *and display it as a graph. Does the error decreases as expected when* k *decreases? Explain.*

Exercise 2.2 *Consider the initial value problem*

$$\frac{dy}{dt} = y + t^2,$$
$$y(0) = 1.$$
(2.25)

(a) Find the exact solution $y(t)$.

(b) Write a computer program that applies the explicit Euler's method to solve (2.25) with $t \in [0, 2]$ for different values of the time step k. For each value $k = 0.001$, $0.002, 0.003, \ldots, 0.05$, your program should find the global error

$$\epsilon(k) = \max_{0 \leq t_n \leq 2} |v(k, t_n) - y(t_n)|.$$

Plot ϵ vs. k and analyze the plot for small k. Is there a linear bound for $\epsilon(k)$? Is there a quadratic bound for $\epsilon(k)$?

2.5   General one-step methods

The one-step methods are a large class of methods generalizing Euler's method that approximate the initial value problem

$$\frac{dy}{dt} = f(y, t),$$

$$y(0) = y_0,$$

$$(2.26)$$

and that compute the approximate solution at time t_{n+1} based on the value of the approximation at t_n only.

Definition 2.9 *A one-step method to approximate (2.26) is a method of the form*

$$v_{n+1} = v_n + k\Phi(v_n, t_n, k), \qquad t_n = nk. \qquad (2.27)$$

The function $\Phi(v, t, k)$ is called the increment function of the method.

The simplest choice of increment function is $\Phi(v, t, k) = f(v, t)$, leading to Euler's explicit method.

Definition 2.10 *A method (2.27) is called locally accurate of order $p + 1$ if*

$$y_{n+1} = y_n + k\Phi(y_n, t_n, k) + \mathcal{O}(k^{p+1}), \qquad (2.28)$$

where $y_n = y(t_n)$ is the exact solution $y(t)$ of (2.26) restricted to the grid.

In other words, the method (2.27) is locally accurate of $\mathcal{O}(k^{p+1})$ if it is obtained from a relation (2.28) by neglecting a term of $\mathcal{O}(k^{p+1})$.

2.6   How to test the correctness of a program

Consider the initial value problem

$$\frac{dy}{dt} = \lambda y + F(t),$$

$$y(0) = y_0,$$

$$(2.29)$$

and approximate it by the forward Euler method

$$v(t + k) = (1 + \lambda k)v(t) + F(t),$$
$$v(0) = y_0.$$

At this point we want to introduce the *solution expansion* (for further details, see Section A.3), one can prove that

$$v(t) = y(t) + k\varphi_1(t) + k^2\varphi_2(t) + \mathcal{O}(k^3). \tag{2.30}$$

Here $y(t)$ is the exact solution of (2.29) and φ_1 and φ_2 are smooth functions of t that do not depend on k.

Assume that we have written a program that carries out Euler's method and we want to test if our program is correct. One cannot overemphasize the importance of such tests. Any experienced programmer distrusts programs and tests them over and over and, for complicated programs, finds errors even after having used them many times.

Test 1 Calculate the numerical solution for a given step size k, resulting in $v^{(1)} = v^{(1)}(t, k)$. Now calculate the solution again but this time with the step size $k/2$, resulting in the solution $v^{(2)} = v^{(2)}(t, k/2)$. By (2.30),

$$v^{(1)}(t, k) - y(t) = k\varphi_1(t) + k^2\varphi_2(t) + \mathcal{O}(k^3), \tag{2.31}$$

$$v^{(2)}(t, k/2) - y(t) = \frac{k}{2}\varphi_1(t) + \frac{k^2}{4}\varphi_2(t) + \mathcal{O}\left(\frac{k^3}{8}\right). \tag{2.32}$$

Therefore, the *precision quotient*,

$$\begin{aligned}
Q(t) &= \frac{v^{(1)}(t, k) - y(t)}{v^{(2)}(t, k/2) - y(t)} \\
&= \frac{k\varphi_1(t) + k^2\varphi_2(t) + \mathcal{O}(k^3)}{(k/2)\varphi_1(t) + (k^4/4)\varphi_2(t) + \mathcal{O}(k^3/8)} \\
&= \frac{2 + 2k(\varphi_2(t)/\varphi_1(t)) + \mathcal{O}(k^2)}{1 + (k/2)(\varphi_2(t)/\varphi_1(t)) + \mathcal{O}(k^2)} \\
&= 2 + \mathcal{O}(k).
\end{aligned} \tag{2.33}$$

Here we have assumed that $\varphi_1(t)$ is not zero. Assume that we have calculated the solution $y(t)$ of (2.29) analytically. Then we can evaluate $Q(t)$ for all $t = nk$, $n = 1, 2, \ldots$ and we can test if $Q(t) \sim 2$. In practice, this test is rather reliable. However, it can happen that the test fails although the program is correct. This can happen for two reasons:

(a) The k used is too large and $k^2\varphi_2(t) + \mathcal{O}(k^3)$ is of the same size as $k\varphi_1(t)$. If, for example, we can neglect the terms $\mathcal{O}(k^3)$ and $k\varphi_1 = k^2\varphi_2$, then $(k/2)\varphi_1 = (k^2/2)\varphi_2$ and, therefore,

$$Q(t) \sim \frac{2 + 2}{1 + 1/2} = \frac{8}{3}. \tag{2.34}$$

The remedy is to alter the test to accommodate smaller k.

(b) One applies the test at a point t where $\varphi_1(t) = 0$ or where $|\varphi_1(t)|$ is very small. If $\varphi_1(t) = 0$ and $\varphi_2(t) \neq 0$, we have for small k:

$$Q(t) \sim 4.$$

Therefore, one applies the test at many points t. If, $Q(t) \sim 2$ at most points and $Q(t) > 2$ at some points, one can be satisfied that the program has passed the test.

For equations more complicated than (2.29), it is often difficult to solve the initial value problem analytically. Let us explain how one proceeds in such cases. Consider the general problem

$$\frac{dy}{dt} = f(y, t),$$
$$y(0) = y_0. \tag{2.35}$$

The function

$$w = y_0 \cos(t)$$

will usually not satisfy (2.35). However, w is a solution of

$$\frac{dw}{dt} = f(w, t) + G(t),$$
$$w(0) = y_0, \tag{2.36}$$

where

$$G(t) = -y_0 \sin(t) + f(y_0 \cos(t), t).$$

We apply our program to solve the new equation (2.36) and can perform the test. An expansion of the form (2.30) is still valid if $f(y, t)$ is a smooth function, so let us assume this fact for now. If the program solves (2.36) correctly, there is a good chance that it is still correct when we choose $G(t) = 0$, that is, when we solve the original equation (2.35).

Test 2   There is another way to test the program which does not require knowledge of an exact solution $y(t)$. Besides the computations with step sizes k and $k/2$, we perform a third calculation with step size $k/4$ and determine $v^{(3)}(t, k/4)$. We obtain

$$v^{(3)}(t, k/4) - y(t) = \frac{k}{4}\varphi_1(t) + \frac{k^2}{16}\varphi_2(t) + \mathcal{O}(k^3). \tag{2.37}$$

Subtracting the expansions (2.32) from (2.31) and, similarly, subtracting (2.37) from (2.32), we obtain

$$v^{(1)}(t, k) - v^{(2)}(t, k/2) = \frac{k}{2}\varphi_1(t) + \frac{3k^2}{4}\varphi_2(t) + \mathcal{O}(k^3),$$
$$v^{(2)}(t, k/2) - v^{(3)}(t, k/4) = \frac{k}{4}\varphi_1(t) + \frac{3k^2}{16}\varphi_2(t) + \mathcal{O}(k^3).$$

We can now calculate

$$\tilde{Q}(t) := \frac{v^{(1)}(t,k) - v^{(2)}(t,k/2)}{v^{(2)}(t,k/2) - v^{(3)}(t,k/4)} = 2 + \mathcal{O}(k). \qquad (2.38)$$

The new precision quotient $\tilde{Q}(t)$ is determined completely by our numerical approximations. If $\tilde{Q}(t) \sim 2$, our program passes the test and has a good chance to be correct. Again, as in Test 1, one has to be careful to choose k small enough. Also, if $|\varphi_1(t)|$ is very small, the test may fail at t even though the program is correct.

Exercise 2.3 *Consider problem (2.29) and an approximation by any one-step method accurate of order p. Show that both precision quotients Q and $\tilde{Q}$ defined in (2.33) and (2.38) should approximate, for small k, the value 2^p at most points.*

Exercise 2.4 *Following Exercise 2.1, modify your program to compute the precision quotient*

$$Q(t) = \frac{v^{(1)}(t) - y(t)}{v^{(2)}(t) - y(t)},$$

where $v^{(1)}(t)$ is the numerical solution computed with step size k and $v^{(2)}(t)$ is the solution computed with step size $k/2$. Show two plots of $Q(t)$, the first computed using $k = 0.01$ and the second, using $k = 0.001$.

2.7 Extrapolation

By combining numerical solutions of a particular method, one can build a higher-order method. This technique is generally known as *extrapolation*.

Consider the initial value problem (2.35) as before. We can use the three computations, with step sizes k, $k/2$, and $k/4$, to improve the accuracy of our approximation. To simplify notation, let us drop the arguments in equations (2.31), (2.32), and (2.37). Then we have

$$y - v^{(1)} = k\varphi_1 + k^2\varphi_2 + \mathcal{O}(k^3), \qquad (2.39)$$

$$y - v^{(2)} = \frac{k}{2}\varphi_1 + \frac{k^2}{4}\varphi_2 + \mathcal{O}(k^3), \qquad (2.40)$$

$$y - v^{(3)} = \frac{k}{4}\varphi_1 + \frac{k^2}{16}\varphi_2 + \mathcal{O}(k^3). \qquad (2.41)$$

Multiply (2.40) by 2 and then subtract (2.39) to obtain

$$y - 2v^{(2)} + v^{(1)} = -\frac{k^2}{2}\varphi_2 + \mathcal{O}(k^3). \qquad (2.42)$$

Similarly, multiply (2.41) by 2 and then subtract from (2.40) to obtain

$$y - 2v^{(3)} + v^{(2)} = -\frac{k^2}{8}\varphi_2 + \mathcal{O}(k^3). \qquad (2.43)$$

The last two equations say that $2v^{(2)} - v^{(1)}$ and $2v^{(3)} - v^{(2)}$ are both order $\mathcal{O}(k^2)$ approximations to $y = y(t)$. Furthermore, if we multiply (2.43) by 4 and then subtract (2.42), we obtain

$$3y + 4(-2v^{(3)} + v^{(2)}) - (-2v^{(2)} + v^{(1)}) = \mathcal{O}(k^3).$$

This says that the combination

$$\frac{1}{3}\left(8v^{(3)} - 6v^{(2)} + v^{(1)}\right)$$

is an $\mathcal{O}(k^3)$ approximation to $y = y(t)$.

Exercise 2.5 *Define a precision quotient $Q(t)$ for the extrapolated solution built above $w(t, k) = (8v^{(3)} - 6v^{(2)} + v^{(1)})/3$ and show that $Q(t) \simeq 2^3 + \mathcal{O}(k)$.*

CHAPTER 3

HIGHER-ORDER METHODS

In this chapter we introduce one-step numerical methods that have been developed to compute higher-precision approximations of the exact solutions of ordinary differential equations. We introduce the second-order Taylor method, the improved Euler's method, and a whole class of methods called Runge-Kutta methods. At the end we generalize the concepts of stability region and truncation error introduced for Euler's method in Chapter 2.

3.1 Second-order Taylor method

To begin with, consider a differential equation

$$\frac{dy}{dt} = \lambda y + F(t), \qquad (3.1)$$

where $F(t)$ is a smooth function. We want to derive a more accurate approximation than the one obtained from the explicit Euler method. If we think of the grid and the notation for grid functions introduced in Section 2.1, we have, by Taylor expansion,

$$y_{n+1} = y_n + k\left(\frac{dy}{dt}\right)_n + \frac{k^2}{2}\left(\frac{d^2y}{dt^2}\right)_n + \frac{k^3}{6}\left(\frac{d^3y}{dt^3}\right)_n + \mathcal{O}(k^4). \qquad (3.2)$$

Introduction to Numerical Methods for Time Dependent Differential Equations, First Edition.
By Heinz-O. Kreiss and Omar E. Ortiz. Copyright © 2014 John Wiley & Sons, Inc.

Now we can use the differential equation to express dy/dt and d^2y/dt^2 in terms of y, F, and dF/dt. This is obvious for dy/dt. Furthermore, the function d^2y/dt^2 is obtained by differentiating the differential equation,

$$\frac{d^2y}{dt^2} = \lambda\frac{dy}{dt} + \frac{dF}{dt} = \lambda^2 y + \lambda F + \frac{dF}{dt}.$$

Therefore, we can write (3.2) as

$$y_{n+1} = \left(1 + \lambda k + \frac{\lambda^2 k^2}{2}\right)y_n + kG_n + k^3 R_n,$$

where

$$G_n = F_n + \frac{k}{2}\left(\lambda F_n + \left(\frac{dF}{dt}\right)_n\right), \quad R_n = \frac{1}{6}\left(\frac{d^3y}{dt^3}\right)_n + \mathcal{O}(k).$$

Neglecting terms of order $\mathcal{O}(k^3)$, we obtain

$$v_{n+1} = \left(1 + \lambda k + \frac{\lambda^2 k^2}{2}\right)v_n + kG_n. \tag{3.3}$$

This is an example of the second-order Taylor method. By construction, the local accuracy of the method is $\mathcal{O}(k^3)$, and the global error in a finite time interval (after integrating $\simeq 1/k$ steps) is $\mathcal{O}(k^2)$.

It is easy to generalize the idea to an initial value problem

$$\begin{aligned}\frac{dy}{dt} &= f(y,t), \\ y(t=0) &= y_0.\end{aligned} \tag{3.4}$$

We have

$$\frac{d^2y}{dt^2} = \frac{\partial f}{\partial y}\frac{dy}{dt} + \frac{\partial f}{\partial t} = \frac{\partial f}{\partial y}f + \frac{\partial f}{\partial t}$$

and obtain

Definition 3.1 *The second-order Taylor method for the initial value problem* (3.4) *is the one-step method given by*

$$v_0 = y_0,$$

$$v_{n+1} = v_n + kf(v_n, t_n) + \frac{k^2}{2}\frac{\partial f}{\partial y}(v_n, t_n)f(v_n, t_n) + \frac{k^2}{2}\frac{\partial f}{\partial t}(v_n, t_n)$$

for $n = 0, 1, 2, \ldots$.

Using higher-order Taylor expansions one can—in principle—obtain methods of a very high order of precision. A disadvantage of this approach is that the partial derivatives of $f(y,t)$ are needed analytically. For example, the second-order Taylor method requires the calculation of $\partial f/\partial y$ and $\partial f/\partial t$. In real situations these analytical derivatives may be difficult to obtain, the expressions obtained complicated, and the evaluation of these derivatives numerically costly, thus making the method inefficient.

It is remarkable that one can obtain higher-order methods that do not use derivatives of $f(y, t)$, only evaluations of f itself. A simple example is given in the next section.

3.2   Improved Euler's method

Definition 3.2 *The improved Euler's method for the initial value problem* (3.4) *is given by*

$$v_0 = y_0$$

and

$$
\begin{aligned}
\tilde{v}_{n+1} &= v_n + k f(v_n, t_n), \\
v_{n+1} &= v_n + k f\left(\frac{1}{2}(v_n + \tilde{v}_{n+1}), t_n + \frac{k}{2}\right)
\end{aligned}
\tag{3.5}
$$

for $n = 0, 1, 2, 3, \ldots$.

In this method any time step $v_n \to v_{n+1}$ is composed by two substeps, each of which requires one evaluation of f. The first substep calculates a preliminary new value $\tilde{v}_{n+1}$ by the explicit Euler method. The second substep computes the new value v_{n+1} by using v_n and $\tilde{v}_{n+1}$.

In terms of the increment function Φ, the improved Euler's method is obtained for

$$\Phi(v, t, k) = f\left(v + \frac{k}{2} f(v, t), t + \frac{k}{2}\right).$$

It is interesting to compare this method with the second-order Taylor method. Applying the improved Euler's method to the example

$$f(y, t) = \lambda y + F(t),$$

we obtain

$$
\begin{aligned}
\tilde{v}_{n+1} &= v_n + \lambda k v_n + k F_n = (1 + \lambda k) v_n + k F_n \\
v_{n+1} &= v_n + k\left(\lambda \frac{1}{2}(v_n + (1 + \lambda k) v_n + k F_n) + F_{n+1/2}\right) \\
&= v_n + k\left(\lambda v_n + \frac{\lambda^2 k}{2} v_n + \frac{\lambda k}{2} F_n + \frac{k}{2}\left(\frac{dF}{dt}\right)_n + \mathcal{O}(k^2)\right) \\
&= \left(1 + \lambda k + \frac{\lambda^2 k^2}{2}\right) v_n + k\left(F_n + \frac{\lambda k}{2} F_n + \frac{k}{2}\left(\frac{dF}{dt}\right)_n\right) + \mathcal{O}(k^3).
\end{aligned}
$$

Comparing with (3.3), we see that the improved Euler's method is coincident up to order k^2 with the second-order Taylor method. The difference between the two methods is locally of order k^3.

3.3 Accuracy of the solution computed

We describe here a procedure that allows us to control the global error of our numerical approximation within a time interval of interest. Consider the initial value problem

$$\frac{dy}{dt} = f(y, t),$$
$$y(0) = y_0,$$

(3.6)

where $f(y, t)$ is a smooth function and approximate it by the explicit Euler method

$$v(t + k) = v(t) + kf(v(t), t),$$
$$v(0) = y_0.$$

We compute two solutions: solution $v^{(1)}$ using step size k and solution $v^{(2)}$ using step size $k/2$. Both solutions satisfy the expansions (see Section A.3)

$$v^{(1)}(nk) = y(nk) + k\varphi_1(nk) + k^2\varphi_2(nk) + \mathcal{O}(k^3),$$
$$v^{(2)}\big((2n)(k/2)\big) = y(nk) + \frac{k}{2}\varphi_1(nk) + \frac{k^2}{4}\varphi_2(nk) + \mathcal{O}(k^3),$$

where $y(t)$ is the exact solution of (3.6). From the second expansion and taking the difference between them, we get

$$v^{(2)}\big((2n)(k/2)\big) - y(nk) = \frac{k}{2}\varphi_1(nk) + \mathcal{O}(k^2),$$
$$v^{(1)}(nk) - v^{(2)}\big((2n)(k/2)\big) = \frac{k}{2}\varphi_1(t) + \mathcal{O}(k^2),$$

which means that to leading order the error of $v^{(2)}$ [i.e., the difference between $v^{(2)}$ and the exact solution $y(t)$], equals the difference between $v^{(1)}$ and $v^{(2)}$.

We can use this information in the following way. Let E denote an error tolerance, the maximum acceptable global error for our numerical approximation. We compute $v^{(1)}(nk)$ and $v^{(2)}\big((2n)(k/2)\big)$ in the largest time interval $0 \le t = nk \le T$ in which

$$|v^{(1)}(nk) - v^{(2)}\big((2n)(k/2)\big)| \le E.$$

(3.7)

If the interval $[0, T]$ does not cover the time interval in which we are interested, we redo the calculations with a smaller value of k. If, after some refinements, the estimate (3.7) holds in the interval of interest, we can expect that $v^{(2)}$ approximates the true solution y with an error $\simeq E$.

In the following we try to analyze the process for a simple model problem. Consider the initial value problem

$$\frac{dy}{dt} = \lambda y - e^{-t},$$
$$y(0) = \frac{1}{1 + \lambda},$$

(3.8)

where $\lambda \neq -1$. The exact solution is $y(t) = e^{-t}/(1+\lambda)$. We approximate using the Euler forward method,

$$v\big((n+1)k\big) = (1 + \lambda k)v(nk) - ke^{-nk}, \tag{3.9}$$

$$v(0) = \frac{1}{1+\lambda}.$$

In Chapter 2 we solved such difference equations. The solution is

$$v(nk) = v_P(nk) + v_H(nk), \tag{3.10}$$

where

$$v_P(nk) = \frac{k}{1 + \lambda k - e^{-k}} e^{-nk} = \left(\frac{1}{1+\lambda} + \frac{k}{2(1+\lambda)^2} + \mathcal{O}(k^2) \right) e^{-nk},$$

$$v_H(nk) = (1 + \lambda k)^n v_H(0),$$

with

$$v_H(0) = \frac{1}{1+\lambda} - v_P(0) = -\frac{k}{2(1+\lambda)^2} + \mathcal{O}(k^2).$$

Now suppose thaht we compute the solution of Euler's method for time steps k and $k/2$ and consider the difference

$$v^{(2)}\big((2n)(k/2)\big) - v^{(1)}(nk) = \Delta_P + \Delta_H,$$

where

$$\begin{aligned} \Delta_P &= v_P^{(2)}\big((2n)(k/2)\big) - v_P^{(1)}(nk), \\ \Delta_H &= v_H^{(2)}\big((2n)(k/2)\big) - v_H^{(1)}(nk). \end{aligned} \tag{3.11}$$

Our formula for $v_P(nk)$ yields

$$|\Delta_P| = \left| \frac{k}{4(1+\lambda)^2} + \mathcal{O}(k^2) \right| e^{-nk}.$$

Therefore, for sufficiently small k,

$$|\Delta_P| \leq \tfrac{1}{2}E \quad \text{for} \quad 0 \leq nk < \infty.$$

On the other hand, the formula for $v_H(t)$ yields

$$|\Delta_H| \leq \left| \frac{k}{4(1+\lambda)^2} + \mathcal{O}(k^2) \right| e^{\operatorname{Re}\lambda(1+\mathcal{O}(\lambda k))nk}.$$

Therefore, the behavior of $|\Delta_H|$ depends on the stability characteristics of the ODE given. If the problem is stable (i.e., $\operatorname{Re}\lambda \leq 0$), the exponential factor is bounded by 1 at all times and we have, for sufficiently small k,

$$|\Delta_H| \leq \tfrac{1}{2}E \quad \text{for} \quad 0 \leq nk < \infty.$$

Thus, if the problem is stable, we can approximate the solution $y(t)$ with a fixed, sufficiently small k. We can expect the error bound E to hold for all times.

If the problem is unstable (i.e., $\operatorname{Re}\lambda > 0$), then

$$|\Delta_H| \leq \tfrac{1}{2}E \quad \text{for} \quad t = nk \leq \tfrac{1}{\operatorname{Re}\lambda + \mathcal{O}(k)} \log_e\left(\tfrac{4(1+\lambda)^2 E}{k} + \mathcal{O}(k)\right)$$

only. Thus the time interval in which the error bound is satisfied increases only in proportion to $\log_e(1/k)$.

Exercise 3.1 *Modify the example above for the case in which the equation is stable (i.e., $\operatorname{Re}\lambda \leq 0$), but the forcing in the equation is exponentially divergent (i.e., e^t instead of e^{-t}). Can we have a bound for the global error for all times?*

Exercise 3.2 *Modify the argument at the beginning of this section to show that, if instead of the explicit Euler method to approximate the solution of (3.1) one uses a one-step method accurate of order p, the corresponding result can be written as follows.*

For a given error tolerance E, neglecting terms of order k^{p+1}, if

$$|v^{(1)}(nk) - v^{(2)}((2n)(k/2))| \leq (2^p - 1)E \quad \text{for} \quad nk \in [0, T],$$

then

$$|v^{(2)}((2n)(k/2)) - y(nk)| \leq E \quad \text{for} \quad nk \in [0, T],$$

where, as before, $y(t)$ denotes the exact solution to the equation.

We present here an example that uses the global error control as described in this section. Assume that we want to solve the initial value problem

$$\begin{aligned} \frac{dy}{dt} &= y - y^3 + \sin(t), \\ y(0) &= 0, \end{aligned} \tag{3.12}$$

for $t \in [0, 10]$, and we want to make sure that the global error of our solution is smaller that $E = 10^{-4}$ during the entire time interval.

Assume also that we decide to approximate the solution with the improved Euler method. Then we write code that implements the improved Euler's method for (3.12) and computes two solutions: the first solution $v^{(1)}$ using step size k and the second solution $v^{(2)}$ using step size $k/2$. Our code stops computing either when the difference $|v^{(1)}(nk) - v^{(2)}((2n)(k/2))|$ reaches the value $(2^2 - 1)E = 3E$, or when $t - 10$. We start using the step size $k = 0.1$ and see that we do not reach the desired accuracy during the entire time interval of interest. We need to do two refinements, dividing the step size by 2 to reach the accuracy desired. Table 3.1 shows the results, and Figure 3.1 shows the solution $v^{(2)}$ and the error for the three runs.

Exercise 3.3 *Consider the initial value problem for a forced pendulum*

$$\frac{d^2\Theta}{dt^2} + \sin\Theta = A\cos(\omega t),$$

$$\Theta(0) = 0, \quad \frac{d\Theta}{dt}(0) = 0.$$

Table 3.1   Maximum values of $|v^{(1)}(nk) - v^{(2)}((2n)(k/2))|$ during computation time with improved Euler approximation of (3.12).

k	$k/2$	t_f	$\max |v^{(1)} - v^{(2)}|$
0.1000	0.0500	0.30	4.43×10^{-4}
0.0500	0.0250	0.65	3.01×10^{-4}
0.0250	0.0125	10.00	8.98×10^{-5}

Reducing this equation to a first-order system with variables $\Theta(t)$ and $d\Theta/dt$, the improved Euler method reads, given the corresponding grid functions u and v, respectively

$$\tilde{u}_{n+1} = u_n + kv_n,$$
$$\tilde{v}_{n+1} = -k\sin(u_n) + kA\cos(\omega t_n),$$
$$u_{n+1} = u_n + \tfrac{1}{2}k(v_n + \tilde{v}_{n+1}),$$
$$v_{n+1} = v_n - k\sin((u_n + \tilde{u}_{n+1})/2) + kA\cos(\omega(t_n + k/2)),$$

with initial data $u_0 = 0$, and $v_0 = 0$.

(a) *Write a computer program that implements this method and solve the problem using $A = 1$ and $\omega = 1$. Let $u^{(1)}$, $u^{(2)}$, and $u^{(3)}$ be the solutions corresponding to k, $k/2$, and $k/4$, respectively. We want to make sure that $|u^{(2)}(t) - \Theta(t)| \leq 0.3 \times 10^{-1}$ when $t \in [0, T]$. Determine with your program the time T one obtains when using $k = 0.1$.*

(b) *Plot, in the interval $[0, T]$, the solutions $u^{(1)}$ and $u^{(2)}$ superimposed.*

(c) *Compute and plot the precision quotient*

$$Q(t_n) = \frac{u^{(2)}(t_n) - u^{(1)}(t_n)}{u^{(3)}(t_n) - u^{(2)}(t_n)}$$

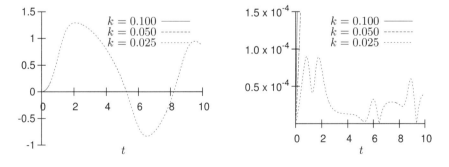

Figure 3.1   Improved Euler approximations to the solution of (3.12) with three time steps $k/2$ (left plot) and estimations of the global error (right plot).

in the time interval $[0, T]$. Did you obtain the result expected?
   (d) Repeat the previous items with $k = 10^{-2}$ and $k = 10^{-3}$.

3.4   Runge-Kutta methods

In Section 2.1 we used Taylor expansion to construct a second-order accurate method for the ordinary differential equation

$$\frac{dy}{dt} = f(y, t), \tag{3.13}$$

where $f(y, t)$ is smooth. In principle, the procedure can be used to derive methods that have any prescribed order, no matter how high. The reason is simple. For any smooth function $y(t)$ we can write the Taylor expansion

$$y(t_n + k) = y(t_n) + \sum_{j=1}^{p} \frac{k^j}{j!} \frac{d^j y(t_n)}{dt^j} + \mathcal{O}(k^{p+1}). \tag{3.14}$$

Then, if $y(t)$ solves (3.13), we can use the differential equation (3.14) to express the time derivatives $d^j y/dt^j$ as functions $f^{(j)}(y, t)$, which are determined by $f(y, t)$ and its partial derivatives of order smaller that or equal to $j - 1$. For example, if we define $f^{(1)}(y, t) = f(y, t)$, equation (3.13) gives

$$\frac{dy}{dt} = f^{(1)}(y, t)$$

and differentiating the equation,

$$\frac{d^2 y}{dt^2} = \frac{d}{dt} f(y, t) = \frac{\partial f}{\partial y}(y, t)\frac{dy}{dt} + \frac{\partial f}{\partial t}(y, t)$$
$$= \frac{\partial f}{\partial y}(y, t)f(y, t) + \frac{\partial f}{\partial t}(y, t).$$

Therefore, we define

$$f^{(2)}(y, t) = \frac{\partial f}{\partial y}(y, t)f(y, t) + \frac{\partial f}{\partial t}(y, t). \tag{3.15}$$

Differentiating once more, one finds that

$$f^{(3)}(y, t) = \frac{\partial^2 f}{\partial y^2}(y, t)f^2(y, t) + 2\frac{\partial^2 f}{\partial t \partial y}(y, t)f(y, t) + \left(\frac{\partial f}{\partial y}(y, t)\right)^2 f(y, t)$$
$$+ \frac{\partial f}{\partial y}(y, t)\frac{\partial f}{\partial t}(y, t) + \frac{\partial^2 f}{\partial t^2}(y, t), \tag{3.16}$$

and the process can be continued.

Neglecting terms of order k^{p+1} in (3.14), we obtain a method that is accurate of order p.

Definition 3.3 *The Taylor method of order p for equation (3.13) is given by*

$$v_{n+1} = v_n + \sum_{j=1}^{p} \frac{k^j}{j!} f^{(j)}(v_n, t_n),$$

where the functions $f^{(j)}(y, t)$ are defined as explained above.

High-order Taylor methods are extremely complicated and not often used in practice. In some applications one does not have an analytic expression of $f(y, t)$ and then cannot compute the functions $f^{(j)}(y, t)$ analytically. In other applications one does have the functions $f^{(j)}(y, t)$ analytically, but their expressions are very complicated and, even worst, computationally too costly to evaluate.

It is the beauty of *Runge-Kutta methods* to avoid any differentiation of $f(y, t)$. All that is required are evaluations of $f(y, t)$ at judiciously chosen points. To make this idea clear we derive the simplest Runge-Kutta methods, which are second-order accurate, and then generalize to higher order.

Assume that $y(t)$ is the solution of (3.13) with initial data $y(t = 0) = y_0$. Making $f^{(1)}$ and $f^{(2)}$ explicit in (3.14), we obtain

$$y(t + k) = y(t) + kf(y, t) + \frac{k^2}{2}\left(\frac{\partial f}{\partial y}(y, t)f(y, t) + \frac{\partial f}{\partial t}(y, t)\right) + \mathcal{O}(k^3)$$

$$= y(t) + k\left(f(y, t) + \frac{\partial f}{\partial y}(y, t)\frac{k}{2}f(y, t) + \frac{\partial f}{\partial t}(y, t)\frac{k}{2}\right) + \mathcal{O}(k^2). \quad (3.17)$$

The key idea is to recognize the terms in parentheses as terms of a two-variable Taylor expansion of $f(y, t)$:

$$f(y + a, t + b) = f(y, t) + \frac{\partial f}{\partial y}(y, t)a + \frac{\partial f}{\partial t}(y, t)b + \frac{\partial^2 f}{\partial y^2}(y, t)a^2$$

$$+ 2\frac{\partial^2 f}{\partial t \partial y}(y, t)ab + \frac{\partial^2 f}{\partial t^2}(y, t)b^2 + \cdots \quad (3.18)$$

Choosing $a = kf(y, t)/2$ and $b = k/2$ in (3.18) and inserting into (3.17), we get

$$y(t + k) = y(t) + kf(y, t) + kf(y + kf(y, t)/2, t + k/2) + \mathcal{O}(k^3). \quad (3.19)$$

Thus, neglecting terms of order k^3 we obtain a method that is second-order accurate and that does not require evaluations of any derivatives of $f(y, t)$ but only two "nested" evaluations of $f(y, t)$. This method can be written as

$$\begin{aligned}
v_0 &= y_0, \\
v_{n+1} &= v_n + k(q_1 + q_2), \qquad n = 0, 1, 2, 3, \ldots, \\
q_1 &= f(v_n, t_n), \\
q_2 &= f\left(v_n + \tfrac{1}{2}kq_1, t_n + \tfrac{1}{2}k\right).
\end{aligned} \quad (3.20)$$

It is direct to check that the method we just obtained is simply the improved Euler's method introduced in Section 3.2.

The procedure just described, starting at (3.17) and leading to the method (3.20), is not unique. For example, we can rewrite (3.17) as

$$y(t+k) = y(t) + \frac{k}{2}f(y,t) + \frac{k}{2}\left(f(y,t) + \frac{\partial f}{\partial y}(y,t)kf(y,t) + k\frac{\partial f}{\partial t}(y,t)\right) + \mathcal{O}(k^3),$$

and then, using (3.18) with $a = kf(y,t)$ and $b = k$, we get

$$y(t+k) = y(t) + \frac{k}{2}f(y,t) + \frac{k}{2}f(y + kf(y,t), t + k) + \mathcal{O}(k^3).$$

Again, neglecting terms of order k^3, we get another second-order accurate method that can be written as

$$
\begin{aligned}
v_0 &= y_0, \\
v_{n+1} &= v_n + \frac{k}{2}(q_1 + q_2), \qquad n = 0, 1, 2, 3, \ldots, \\
q_1 &= f(v_n, t_n), \\
q_2 &= f\left(v_n + kq_1, t_n + k\right).
\end{aligned}
\tag{3.21}
$$

This method is known in the literature as the *method of Heun*.

The improved Euler method and the method of Heun are simply particular examples of Runge-Kutta methods of order 2.

Definition 3.4 *Runge-Kutta methods of order 2 for the initial value problem for equation (3.13) with initial data* $y(t = 0) = y_0$ *are*

$$
\begin{aligned}
v_0 &= y_0, \\
v_{n+1} &= v_n + k(\alpha_1 q_1 + \alpha_2 q_2), \qquad n = 0, 1, 2, 3, \ldots, \\
q_1 &= f(v_n, t_n), \\
q_2 &= f(v_n + \beta_{21} k q_1, t_n + \gamma_2 k),
\end{aligned}
$$

where the coefficients $\alpha_1, \alpha_2, \beta_{21},$ *and* γ_2 *satisfy*

$$\alpha_1 + \alpha_2 = 1, \quad \alpha_2 \beta_{21} = \tfrac{1}{2}, \quad \alpha_2 \gamma_2 = \tfrac{1}{2}.$$

Runge-Kutta methods of order 2 are one-step methods that require two nested evaluations of the function $f(y,t)$ and no evaluation of any derivative of f. The incremental function of these methods is

$$\Phi(v_n, t_n, k) = \alpha_1 f(v_n, t_n) + \alpha_2 f(v_n + \beta k f(v_n, t_n), t_n + \gamma k).$$

The four coefficients $\alpha_1, \alpha_2, \beta_{21},$ and γ_2 that define a particular method need to satisfy only three nonlinear algebraic equations. As seen earlier different solutions to these algebraic equations can be found, leading to different Runge-Kutta methods of order 2. The nonuniqueness of these solutions is a very important property that holds for all Runge-Kutta methods of any order. This freedom in choosing the coefficients that define a Runge-Kutta method can be exploited in various ways, such as minimizing the error for a particular equation or building embedded Runge-Kutta methods useful to control the time step to keep the error under tolerance. In Section 4.3 we present a very simple variable-step-size strategy that adjusts the time step using an estimate of the local error dominant term. More information on step size control and embedded Runge-Kutta methods is available in the literature [4, 8].

Runge-Kutta methods of order p for initial value problem (3.13) with initial data $y(t = 0) = y_0$ are methods of the form

$$v_0 = y_0,$$

$$v_{n+1} = v_n + k \sum_{j=1}^{p} \alpha_j q_j, \qquad n = 0, 1, 2, 3, \ldots,$$

where

$$q_1 = f(v_n, t_n),$$
$$q_2 = f(v_n + \beta_{21} k q_1, t_n + \gamma_2 k),$$
$$\ldots \ldots$$
$$q_p = f(v_n + \beta_{p1} k q_1 + \beta_{p2} k q_2 + \cdots + \beta_{p,p-1} k q_{p-1}, t_n + \gamma_p k),$$

and where the coefficients α_j, β_{jk}, and γ_j satisfy a set of nonlinear algebraic equations so that the method is locally accurate of order $p + 1$. There is no point in writing the set of equations for a general method of order p here. We would rather concentrate on important examples.

It is usual to present a particular Runge-Kutta method of order p by providing a table with the coefficients that define the method. Care should be taken since the table can be presented in different ways. A possible way is

$$
\begin{array}{llllll}
\alpha_1 & 0 & & & & \\
\alpha_2 & \gamma_2 & \beta_{21} & & & \\
\alpha_3 & \gamma_3 & \beta_{31} & \beta_{32} & & \\
\cdots & \cdots & \cdots & \cdots & \cdots & \\
\alpha_p & \gamma_p & \beta_{p1} & \beta_{p2} & \cdots & \beta_{p,p-1}
\end{array}
\qquad (3.22)
$$

For example, the improved Euler and Heun methods are given by

$$
\begin{array}{ll}
1 & 0 \\
& \\
1 & \frac{1}{2} \quad \frac{1}{2}
\end{array}
\qquad \text{and} \qquad
\begin{array}{lll}
\frac{1}{2} & 0 & \\
& & \\
\frac{1}{2} & 1 & 1,
\end{array}
$$

respectively.

By far, the Runge-Kutta method most widely used in applications is the *classical fourth-order Runge-Kutta method* with the table

$$
\begin{array}{c|ccccc}
\frac{1}{6} & 0 \\
\frac{1}{3} & \frac{1}{2} & \frac{1}{2} \\
\frac{1}{3} & \frac{1}{2} & 0 & \frac{1}{2} \\
\frac{1}{6} & 1 & 0 & 0 & 1;
\end{array}
$$

Or explicitly,

Definition 3.5 *The classical fourth-order Runge-Kutta method for the initial value problem* (3.13) *with initial data* $y(t = 0) = y_0$ *is*

$$
v_0 = y_0,
$$
$$
v_{n+1} = v_n + \frac{k}{6}(q_1 + 2q_2 + 2q_3 + q_4), \qquad n = 0, 1, 2, 3, \ldots,
$$

where

$$
q_1 = f(v_n, t_n),
$$
$$
q_2 = f\left(v_n + \frac{k}{2}q_1, t_n + \frac{k}{2}\right),
$$
$$
q_3 = f\left(v_n + \frac{k}{2}q_2, t_n + \frac{k}{2}\right),
$$
$$
q_4 = f(v_n + kq_3, t_n + k).
$$

Exercise 3.4 *Write a code that implements the classical fourth-order Runge-Kutta method to solve problem* (3.12) *for* $t \in [0, 10]$. *Perform runs with time steps* $k = 10^{-2}$, $k/2$, *and* $k/4$. *Plot the solution* $v(t, k/4)$. *Your code should also compute the precision quotient*

$$
Q(t, k) = \frac{v(t, k) - v(t, k/2)}{v(t, k/2) - v(t, k/4)}.
$$

Based on the solution expansion, show that $Q(t)$ *should be close to the value* $2^4 = 16$ *most of the time. Plot the* $Q(t)$ *obtained with your code. Is the result satisfactory?*

3.5   Regions of stability

Consider the initial value problem

$$
\frac{dy}{dt} = f(y, t), \tag{3.23}
$$
$$
y(0) = y_0.
$$

The solutions v_n of all the difference approximations that we have discussed to approximate (3.23) converge, in maximum norm, to the solution $y(t)$ for $k \to 0$. More precisely, for any finite interval $0 \le t \le T$ where the exact solution $y(t)$ exists and is smooth, we have

$$\max_{0 \le t_n \le T} |v_n - y_n| \le Ck^p, \tag{3.24}$$

where $C = C(T)$ is a constant independent of step size k and p is the accuracy order of the method. We have seen that $p = 1$ for the explicit Euler method, $p = 2$ for the improved Euler method, $p = 4$ for the classical Runge-Kutta method, and p depends on the order of expansion for Taylor's method. As for the case of the explicit Euler method, the estimate (3.24) implies that the error converges to zero as $k \to 0$.

We now want to generalize the idea of a stability region, introduced for Euler's method in Chapter 2, to other one-step methods. If one applies any of the one-step difference methods that we have discussed so far to the model problem (2.10), which we repeat here for convenience:

$$\frac{dy}{dt} = \lambda y, \quad \lambda \in \mathbb{R}, \ \operatorname{Re} \lambda \le 0,$$
$$y(0) = 1, \tag{3.25}$$

one obtains

$$v_{n+1} = (1 + Q(\lambda k))v_n,$$
$$v_0 = y_0, \tag{3.26}$$

where $Q(\lambda k)$ is a polynomial in $\mu = \lambda k$, that depends on the method. The stability region of the method is defined by a requirement analogous to (2.12).

Definition 3.6 *Assume that a one-step difference method applied to* (3.25) *yields* (3.26). *Then the stability region of the method consists of all complex numbers μ that satisfy*

$$|1 + Q(\mu)| \le 1.$$

We compute and plot here, as an example, the stability region of the classical Runge-Kutta method. Definition 3.5 applied to model problem (3.5) gives

$$q_1 = \lambda k v_n,$$
$$q_2 = \left(\lambda + \tfrac{1}{2}k\lambda^2\right),$$
$$q_3 = \left(\lambda + \tfrac{1}{2}k\lambda^2 + \tfrac{1}{4}k^2\lambda^3\right)v_n,$$
$$q_4 = \left(\lambda + k\lambda^2 + \tfrac{1}{2}k^2\lambda^3 + \tfrac{1}{4}k^3\lambda^4\right)v_n,$$

so that

$$v_{n+1} = \left(1 + \mu + \tfrac{1}{2}\mu^2 + \tfrac{1}{6}\mu^3 + \tfrac{1}{24}\mu^4\right)v_n = (1 + Q(\mu))v_n,$$

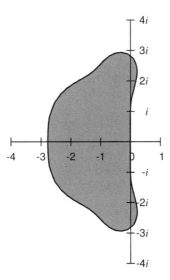

Figure 3.2   The shadowed region is the stability region of the classical fourth-order Runge-Kutta method.

with $\mu = \lambda k$. The stability region is the region in the complex μ-plane that satisfies $|1 + Q(\mu)| \le 1$. If we plot this region in gray, we get Figure 3.2.

By definition, one can determine the stability region of a method by studying how the method behaves when applied to model equation (3.25). However, the stability region is of general importance: If one solves the equation $dy/dt = f(y, t)$, one always tries to choose k so that $\lambda_n k$ lies in the stability region, where $\lambda_n = \partial f(v_n, t_n)/\partial y$. We discuss below why this is reasonable.

Because of its general importance, the stability region has been determined for every numerical method used in practice (see, e.g., [8]).

Exercise 3.5 *Determine and plot the stability region for the improved Euler's method.*

Exercise 3.6 *For the equation $y = i\lambda y$, $\lambda \in \mathbb{R}$, determine the stability interval for the third-order method based on Taylor expansion.*

Exercise 3.7 *Consider the explicit Euler method applied to the equation*

$$\frac{dy}{dt} = \lambda y,$$

where λ is purely imaginary.
   (a) Is it possible to choose step k so that λk belongs to the stability region?
   (b) Modify the Euler method so that

$$v_{n+1} = (1 + \lambda k + \alpha \lambda^2 k^2) v_n.$$

Find $\alpha \in \mathbb{R}$ so that the interval on the imaginary axis of the λk-plane that belongs to the stability region is as large as possible.

3.6 Accuracy and truncation error

We generalize here the concept of truncation error, introduced in Definition 2.4 for Euler's method, to a general one-step method with increment function $\Phi(v, t, k)$.

Definition 3.7 *Consider an initial value problem*

$$\frac{dy}{dt} = f(y, t),$$
$$y(0) = y_0,$$

and approximate it by a one-step method,

$$v_{n+1} = v_n + k\Phi(v_n, t_n, k),$$
$$v_0 = y_0.$$

Let $y(t)$ denote the solution of the initial value problem in some interval $0 \le t \le T$ and substitute $y_n = y(t_n)$ into the difference equations. The truncation error R_n is defined as

$$R_n = \frac{y_{n+1} - y_n}{k} - \Phi(y_n, t_n, k). \tag{3.27}$$

If

$$\max_{0 \le t_n \le T} |R_n| = \mathcal{O}(k^p), \tag{3.28}$$

the method is said to be accurate of order p.

As before, the truncation error R_n depends on the step size k and on the solution $y(t)$, although we usually suppress this in our notation. As a rule, the accuracy order p of a method depends neither on the particular solution $y(t)$ nor on the particular equation $dy/dt = f(y, t)$ that one considers. However, there are exceptions to this rule. For example, explicit Euler's method is a first-order method. However if one applies Euler's method to the trivial equation $dy/dt = 0$, one obtains the exact solution. Thus, in this exceptional case, the approximation is accurate to any order.

Exercise 3.8 *Derive the truncation error for*
    (a) the improved Euler's method
    (b) the method of Heun
applied to the general problem

$$\frac{dy}{dt} = f(y, t),$$
$$y(0) = y_0.$$

In both cases, find an explicit expression for the lower-order term of R_n in terms of f and its derivatives, and show that these methods are indeed accurate of order 2.

Exercise 3.9 *Determine the truncation error for the method introduced in Exercise 3.7 (b). Is the best choice of α from the point of view of accuracy coincident with the best choice of α from the point of view of stability? [compare with your solution to exercise 3.7 (b)]. In a way similar to that used in Theorem 2.7, estimate the global error as a function of α.*

3.7   Difference approximations for unstable problems

In Section 1.3 we discussed the concept of stability. A problem is called unstable if perturbations grow exponentially. A simple example is given by

$$\frac{du}{dt} = \lambda u, \quad \lambda = \lambda_R + i\lambda_I, \quad \lambda_R > 0,$$
$$u(0) = u_0.$$

We approximate it by the explicit Euler method

$$v_{n+1} = (1 + \lambda k)v_n,$$
$$v_0 = u_0.$$

We cannot choose $k > 0$ so that $k\lambda$ belongs to the stability region of the method since

$$|1 + \lambda k|^2 = (1 + \lambda_R k)^2 + \lambda_I^2 k^2 = 1 + 2\lambda_r k + k^2 |\lambda|^2 > 1.$$

In fact, there is no method for which λk belongs to its stability region for $0 < k < k_0$, due to the following facts:

- The solution to the differential equation grows exponentially.

- If λk belongs to the stability region, the solution of the difference approximation is bounded.

- For $k \to 0$, the solution of the difference approximation converges to the solution of the differential equation.

We have to relax our requirement that λk belongs to the stability region.

Definition 3.8 *Approximate $du/dt = \lambda u$, $u(0) = 0$, by the difference approximation*

$$v_{n+1} = Q(\lambda k)v_n,$$
$$v_0 = u_0.$$

We call the approximation conditionally stable if its solution does not grow, in absolute value, faster than the solution of the differential equation, that is,

$$|Q(\lambda k)| \leq e^{\lambda_R k}. \tag{3.29}$$

As an example, we consider the explicit Euler method. In this case

$$|Q(\lambda k)|^2 = (1 + \lambda_R k)^2 + \lambda_I^2 k^2 = 1 + 2\lambda_R k + k^2(\lambda_R^2 + \lambda_I^2).$$

Since

$$(e^{\lambda_R k})^2 > 1 + 2\lambda_R k + 2\lambda_R^2 k^2,$$

the condition (3.29) is satisfied if $|\lambda_I| \leq |\lambda_R|$ and the explicit Euler is conditionally stable.

CHAPTER 4

IMPLICIT EULER METHOD

Some differential equations, such as stiff equations, are difficult to approximate using explicit methods because the stability condition imposes severe restrictions on the time step. It is therefore very useful to introduce a new class of methods, called implicit methods. In this chapter we explain some basic properties of stiff equations by analyzing a simple model equation. Then we introduce the implicit Euler method, which is the most basic implicit method. At the end of the chapter a very simple variable-step-size strategy is explained. This strategy shows the fundamental idea behind more complex strategies used in higher-order methods.

4.1 Stiff equations

Many applications lead to *Stiff differential equations*. A simple example of a stiff equation is

$$\frac{dy}{dt} = -10^3 y + 10^3 \sin(t),$$

$$y(0) = y_0,$$

Introduction to Numerical Methods for Time Dependent Differential Equations, First Edition.   **55**
By Heinz-O. Kreiss and Omar E. Ortiz. Copyright © 2014 John Wiley & Sons, Inc.

which can also be written as

$$\varepsilon \frac{dy}{dt} = -y + \sin(t), \quad \varepsilon = 10^{-3},$$

$$y(0) = y_0. \tag{4.1}$$

This equation can be solved exactly using lemma 1.2. The exact solution is

$$y(t) = e^{-t/\varepsilon}\left(y_0 + \frac{\varepsilon}{1 + \varepsilon^2}\right) + \frac{1}{1 + \varepsilon^2}\left(\sin(t) - \varepsilon \cos(t)\right). \tag{4.2}$$

To gain insight into the behavior of the solution, instead of analyzing the exact solution (4.2), we prefer to use a general procedure to obtain the dominant terms of this solution. This procedure has the advantage that it can be applied to more general problems where the exact solutions are not known.

If we formally set $\varepsilon = 0$ in (4.1), we see that the function $y = \sin(t)$ satisfies the equation. This motivates us to define a new variable y_1 by

$$y = \sin(t) + y_1.$$

For the function $y_1(t)$, one obtains

$$\varepsilon \frac{dy_1}{dt} = -y_1 - \varepsilon \cos(t),$$

$$y_1(0) = y_0. \tag{4.3}$$

Thus, the forcing in the y_1-equation is reduced to order $\mathcal{O}(\varepsilon)$. We repeat the process and define y_2 by

$$y_1 = -\varepsilon \cos(t) + y_2.$$

Then y_2 satisfies

$$\varepsilon \frac{dy_2}{dt} = -y_2 - \varepsilon^2 \sin(t),$$

$$y_2(0) = y_0 + \varepsilon,$$

and the forcing is of order $\mathcal{O}(\varepsilon^2)$. Clearly, we could continue the process and reduce the forcing to even higher order in ε.

Since the initial data in the y_2-equation are $\mathcal{O}(1)$, in general, the function y_2 will not be small. Therefore, we now solve

$$\varepsilon \frac{db}{dt} = -b, \quad b(0) = y_0 + \varepsilon,$$

that is,

$$b(t) = e^{-t/\varepsilon}(y_0 + \varepsilon).$$

Then the function $r(t) = y_2(t) - b(t)$ solves

$$\varepsilon \frac{dr}{dt} = -r - \varepsilon^2 \sin(t),$$

$$r(0) = 0.$$

The function $r(t)$ plays the role of a remainder term: The forcing and the initial data in the r-equation are both small, and therefore we can expect $r(t)$ to be small. Indeed, using the explicit formula (1.9), we obtain

$$r(t) = \varepsilon \int_0^t e^{(t-s)/\varepsilon} \sin(s)\, ds,$$

and can estimate $|r(t)|$,

$$|r(t)| \le \varepsilon \int_0^t e^{-(t-s)/\varepsilon}\, ds = \varepsilon^2 e^{(t-s)/\varepsilon}\Big|_0^t \le \varepsilon^2.$$

To summarize, the solution $y(t)$ of (4.1) has the following representation:

$$\begin{aligned}
y &= \sin(t) + y_1 \\
&= \sin(t) - \varepsilon \cos(t) + y_2 \\
&= \sin(t) - \varepsilon \cos(t) + b(t) + r(t) \\
&= \sin(t) - \varepsilon \cos(t) + e^{-t/\varepsilon}(y_0 + \varepsilon) + r(t) \quad \text{with} \quad |r(t)| \le \varepsilon^2. \qquad (4.4)
\end{aligned}$$

The representation (4.4) provides the leading terms of the *asymptotic expansion* of the exact solution $y(t)$. It shows that except for a small remainder $r(t) = \mathcal{O}(\varepsilon^2)$, the solution of (4.3) consists of a slowly varying part, $\sin(t) - \varepsilon \cos(t)$, and an *initial layer*, $e^{-t/\varepsilon}(y_0 + \varepsilon)$. Figure 4.1 shows a plot of both the exact solution (4.2) and the first three terms of the asymptotic expansion (4.4) [i.e., without the remainder $r(t)$].

Suppose that we try to solve (4.1) numerically by the explicit Euler method. To obtain a reasonable approximation, we must choose the step size k so that $\lambda k = -k/\varepsilon$ belongs to the stability region of the method (i.e., so that $k \le 2\varepsilon$). From an accuracy point of view, such a small step size is adequate in the initial layer, where the solution changes rapidly. Away from the initial layer, one would like to choose

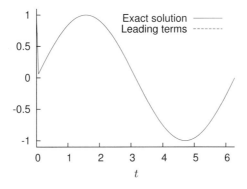

Figure 4.1   The exact solution (4.2) and the leading terms in the asymptotic expansion (4.4) look completely superimposed in this plot.

a larger step size since the solution varies slowly. With explicit Euler, however, a larger step size cannot be chosen. If $k > 2\varepsilon$, the method becomes unstable, which leads to growing rapid oscillations.

4.2   Implicit Euler method

When approximating the solution of (4.1), the possibility of choosing a small time step to solve the initial layer and a bigger time step for the rest of the solution can be implemented for the *implicit Euler method.* Assume that we want to approximate

$$\frac{dy}{dt} = \lambda y + F(t),$$
$$y(0) = y_0;$$

(4.5)

then

Definition 4.1 *The implicit Euler method to approximate* (4.5) *is defined by*

$$v_0 = y_0,$$
$$\frac{v_{n+1} - v_n}{k} = \lambda v_{n+1} + F_{n+1}, \quad n = 0, 1, 2, \ldots.$$

(4.6)

For (4.1) the approximation reads

$$\varepsilon \frac{v_{n+1} - v_n}{k} = -v_{n+1} + F_{n+1}, \quad F_{n+1} = \sin(t_{n+1}) = \sin((n+1)k).$$

To discuss how the approximation behaves, we derive an asymptotic expansion for $v_n = v_n(k, \varepsilon)$ which is similar to the asymptotic expansion of the exact solution $y(t)$ given in (4.4).

First we make a change of variables,

$$v_n = \sin(t_n) + (v_1)_n,$$

and obtain

$$\varepsilon \frac{(v_1)_{n+1} - (v_1)_n}{k} = -(v_1)_{n+1} - \varepsilon D_- \sin(t_{n+1}),$$
$$(v_1)_0 = y_0.$$

Here we denote by D_- the difference operator

$$D_- f(t_n) = \frac{f(t_n) - f(t_{n-1})}{k},$$

so that, by Taylor expansion,

$$D_- \sin(t_{n+1}) = \frac{\sin(t_{n+1}) - \sin(t_n)}{k} = \cos(t_{n+1}) + \mathcal{O}(k).$$

We repeat the process and set

$$(v_1)_n = -\varepsilon D_- \sin(t_n) + (v_2)_n.$$

Then $(v_2)_n$ satisfies

$$\varepsilon \frac{(v_2)_{n+1} - (v_2)_n}{k} = -(v_2)_{n+1} + \varepsilon^2 D_-^2 \sin(t_{n+1}),$$

$$(v_2)_0 = y_0 + \varepsilon D_- \sin(t_0) = y_0 + \varepsilon + \mathcal{O}(k\varepsilon).$$

The equation for $(v_2)_n$ can be written as

$$(v_2)_{n+1} = \frac{\varepsilon}{\varepsilon + k}(v_2)_n + \frac{\varepsilon^2 k}{\varepsilon + k} D_-^2 \sin(t_{n+1}). \tag{4.7}$$

To obtain the *discrete limit layer,* we solve the homogeneous equation

$$\varepsilon \frac{w_{n+1} - w_n}{k} = -w_{n+1}; \quad w_0 = (v_2)_0;$$

that is,

$$w_{n+1} = \frac{\varepsilon}{\varepsilon + k} w_n.$$

Thus,

$$w_n = S(n, 0) w_0,$$

where

$$S(n, m) = \left(\frac{\varepsilon}{\varepsilon + k}\right)^{n-m}$$

is the solution operator. Then if we define

$$(v_2)_n = w_n + \rho_n,$$

ρ_n satisfies the same difference equations as $(v_2)_n$ does, but the initial value is $\rho_0 = 0$. We can now use the discrete version of Duhamel's principle. If we identify ρ_n with e_n and the inhomogeneity in (4.7) with kR_n, we can apply Lemma 2.6. We have

$$\rho_n = \sum_{m=0}^{n-1} \left(\frac{\varepsilon}{\varepsilon + k}\right)^{n-1-m} \frac{\varepsilon^2 k}{\varepsilon + k} D_-^2 \sin(t_{m+1}).$$

By Taylor expansion

$$D_-^2 \sin(t_{m+1}) = -\sin(t_m) + \mathcal{O}(k),$$

so that

$$|D_-^2 \sin(t_{m+1})| \le 1 + \mathcal{O}(k).$$

Therefore,

$$|\rho_n| \le \frac{\varepsilon^2 k}{\varepsilon + k}(1 + \mathcal{O}(k)) \sum_{m=0}^{n-1} \left(\frac{\varepsilon}{\varepsilon + k}\right)^{n-1-m}. \tag{4.8}$$

The last factor on the right-hand side is a finite geometric series [see (A.1)], so that

$$\sum_{m=0}^{n-1}\left(\frac{\varepsilon}{\varepsilon+k}\right)^{n-1-m} = \sum_{j=0}^{n-1}\left(\frac{\varepsilon}{\varepsilon+k}\right)^{j} = \frac{1-[\varepsilon/(\varepsilon+k)]^n}{1-\varepsilon/(\varepsilon+k)}.$$

As $\varepsilon/(\varepsilon+k) < 1$, from (4.8) we have

$$|\rho_n| \le \varepsilon^2(1+\mathcal{O}(k)).$$

Thus, for the complete solution v_n we get

$$v_n = \sin(t_n) - \varepsilon\cos(t_n) + w_n + \varepsilon^2(1+\mathcal{O}(k)). \tag{4.9}$$

Comparing (4.9) with (4.4), we see that the slowly varying part of v_n agrees with the corresponding part of the exact solution, except for terms of order $\mathcal{O}(\varepsilon^2)$. The comparison of the initial layer terms is more subtle. Here we must compare

$$e^{-t_n/\varepsilon} = (e^{-k/\varepsilon})^n \tag{4.10}$$

with the corresponding discrete term

$$\left(\frac{\varepsilon}{\varepsilon+k}\right)^n = \left(\frac{1}{1+k/\varepsilon}\right)^n. \tag{4.11}$$

By Taylor expansion

$$e^{-k/\varepsilon} = \frac{1}{1+k/\varepsilon} + \mathcal{O}\left(\frac{k}{\varepsilon}\right).$$

Thus, if $k/\varepsilon \ll 1$, the two terms (4.10) and (4.11) are close for all $n = 0, 1, \ldots$. On the other hand, if k/ε is not small, the two terms (4.10) and (4.11) are not close to each other for small n, and therefore v_n is not a good approximation for y_n in the initial layer. If k/ε is not small but n is large, terms (4.10) and (4.11) are both close to zero. This implies that v_n is a good approximation for y_n outside the initial layer, even if k/ε is not small.

In Figure 4.2 we have plotted the error $e_n = y_n - v_n$ for $y_0 = 1$, $\varepsilon = 10^{-3}$, $k = 10^{-2}$. The exact solution was given in (4.2), while v_n was computed numerically. Clearly, $k/\varepsilon = 10$ is not small. Nevertheless, as predicted by our analysis, away from the initial layer the error is small.

We now use the truncation error analysis to discuss the error of the implicit Euler method in the more general case (4.5). We assume that $\operatorname{Re}\lambda \le 0$, with $|\lambda|$ not large. By definition the truncation error R_n is obtained when we substitute the exact solution $y(t)$ in the difference equations (4.6). By Taylor expansion,

$$y_n = y_{n+1} - k(y_t)_{n+1} + \frac{k^2}{2}(y_{tt})_{n+1} + \mathcal{O}(k^3),$$

and therefore,

$$\frac{y_{n+1}-y_n}{k} - \lambda y_{n+1} - F_{n+1} = (y_t)_{n+1} - \lambda y_{n+1} - F_{n+1} - \frac{k}{2}(y_{tt})_{n+1} + \mathcal{O}(k^2)$$

$$= -\frac{k}{2}(y_{tt})_{n+1} + \mathcal{O}(k^2)$$

$$= R_n.$$

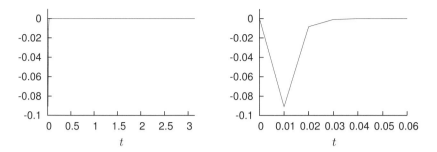

Figure 4.2   Error $e_n = y_n - v_n$ in the interval $t \in [0, 2\pi]$ (left plot), and of the initial layer, $t \in [0, 0.06]$, (right plot).

We can write this in the form

$$y_{n+1} = \frac{1}{1 - \lambda k} y_n + \frac{k}{1 - \lambda k} F_{n+1} + \frac{k}{1 - \lambda k} R_n. \tag{4.12}$$

Let us note that, to leading order, the truncation errors for the explicit and implicit Euler methods are the same: We have $R_n = (k/2)(y_{tt})_n + \mathcal{O}(k^2)$ in both cases. However, the stability region for the two methods is completely different. For the explicit Euler method, as we have seen in Section 2.2, the stability region consisted of all points λk in the complex plane with

$$|1 + \lambda k| \leq 1;$$

that is, the stability region is a disk with radius 1 about the point -1. For the implicit Euler method the amplification factor is $(1 - \lambda k)^{-1}$, and we have

$$\left| \frac{1}{1 - \lambda k} \right| \leq 1,$$

which holds if and only if $|1 - \lambda k| \geq 1$; That is, the stability region consists of all points λk in the complex plane outside the disk of radius 1 about point 1 (see Figure 4.3). In particular, all points λk with $\mathrm{Re}\,\lambda < 0$ are in the stability region, no matter how large one chooses the step size k. One says that for $\mathrm{Re}\,\lambda < 0$, the implicit Euler method is unconditionally stable.

Let us discuss the implications of this good stability property for the error $e_n = y_n - v_n$. Notice that (4.6) can also be written as

$$v_{n+1} = \frac{1}{1 - \lambda k} v_n + \frac{k}{1 - \lambda k} F_{n+1},$$

$$v_0 = y_0. \tag{4.13}$$

Subtracting (4.13) from (4.12), we obtain

$$e_{n+1} = \frac{1}{1 - \lambda k} e_n + \frac{k}{1 - \lambda k} R_n,$$

$$e_0 = 0.$$

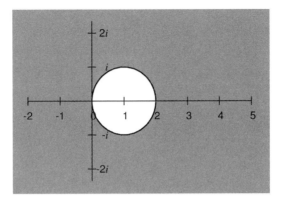

Figure 4.3 The shadowed region is the stability region of the implicit Euler method.

By Lemma 2.6,

$$e_n = \frac{k}{1-\lambda k} \sum_{m=0}^{n-1} S_k(n-1,m)R_m, \quad R_m = \frac{k}{2}(y_{tt})_m + \mathcal{O}(k^2), \quad (4.14)$$

where

$$S_k(n,m) = \frac{1}{(1-\lambda k)^{n-m}}.$$

By assumption we have $\operatorname{Re}\lambda < 0$, and therefore λk always belongs to the stability region. This yields

$$|S_k(n,m)| \leq 1,$$

and the representation (4.14) implies that

$$|e_n| \leq \sum_{m=0}^{n} |kR_m|. \quad (4.15)$$

To leading order, $kR_m \simeq (k^2/2)(y_{tt})_m$, and therefore (4.15) tells us that we should choose k so that $|(k^2)(y_{tt})_m|$ is small everywhere.

What does this say for our example (4.1)? In the example we have, using (4.4),

$$\frac{k^2}{2}\left|\frac{d^2y}{dt^2}\right| \simeq \left|\frac{k^2}{2\varepsilon}e^{-t/\varepsilon}(y_0+\varepsilon) - \frac{k^2}{2}(\sin(t) - \varepsilon\cos(t))\right| \simeq \frac{k^2}{2\varepsilon^2}e^{-t/\varepsilon} + \mathcal{O}(k^2). \quad (4.16)$$

The $\mathcal{O}(k^2)$-term is bounded by Ck^2 with C independent of ε. In the initial layer, where the term $e^{-t/\varepsilon}$ is $\mathcal{O}(1)$, we have to choose k so that $k \ll \varepsilon$ in order to make $(k^2/2)|y_{tt}|$ small. On the other hand, outside the initial layer, where the term $e^{-t/\varepsilon}$ is small, it suffices to choose $k \ll 1$. This requirement is then independent of ε. Our analysis suggests that we work with two different step sizes (see Figure 4.4): First we divide the t-axis into two intervals, $0 \leq t \leq \bar{t}$ and $\bar{t} \leq t \leq \infty$. Here

$$\bar{t} = \bar{t}(\varepsilon) = \mathcal{O}(\varepsilon \log(1/\varepsilon))$$

Figure 4.4   Grid with two different step sizes.

is chosen so that $\varepsilon^{-2}e^{-t_\varepsilon/\varepsilon} \ll 1$. The interval $0 \le t \le \bar{t}$ is viewed as the initial layer, because the exponential term $\varepsilon^2 e^{-t/\varepsilon}$ in (4.16) is small for $t > \bar{t}$. Second, in the interval $0 \le t \le \bar{t}$, we use a step size k with $k/\varepsilon \ll 1$. In the remaining interval, $t \ge \bar{t}$, we can work with a step size k satisfying only $k \ll 1$.

For more complicated problems, which cannot be analyzed as easily, it is a good and common practice to use methods that regulate the step size during computations.

4.3   Simple variable-step-size strategy

Let us describe a simple strategy to construct a variable-step-size implicit Euler method. The idea is to compute, at every time step, an approximation of the leading local error term $(k_n^2/2)(d^2y/dt^2)_n$ $(k_n^2/2)(d^2y/dt^2)_n$ and to regulate the step size so as to keep this term between two given threshold values that we call M and N with $0 < M < N$. More precisely, assume that v_n is the computed numerical value at time t_n and k_n is the current step size. Then we compute the preliminary values $\tilde{v}_{n+1/2}$ and $\tilde{v}_{n+1}$ by using a step size $k_n/2$ once and twice respectively. Using these preliminary values, we compute

$$Q_n = 2(\tilde{v}_{n+1} - 2\tilde{v}_{n+1/2} + v_n),$$

which is simply an approximation of the leading error term at the intermediate point $t_n + k_n/2$. Now:

1. If
$$M \le |Q_n| \le N,$$

   the error is within the values expected and we accept $\tilde{v}_{n+1}$ as the value of our numerical approximation at $t_{n+1} = t_n + k_n$ and set $k_{n+1} = k_n$.

2. If
$$|Q_n| < M,$$

   our code computed $\tilde{v}_{n+1}$ with a precision level higher than that requested. We accept $\tilde{v}_{n+1}$ as the value of our numerical approximation at $t_{n+1} = t_n + k_n$ but try the next step with a larger step size by setting $k_{n+1} = \frac{3}{2}k_n$.

3. If
$$|Q_n| > N,$$

   then the step, and so the error, was too big. We replace k_n by the smaller step size $\frac{2}{3}k_n$, and repeat the computation of $\tilde{v}_{n+1/2}$ and $\tilde{v}_{n+1}$.

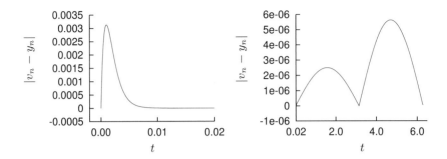

Figure 4.5   Error as a function of time. The error in the initial layer is on the left, and the error in the rest of the interval is on the right. Observe the different scales.

As an example we show the solution to our model problem (4.1) for $t \in [0, 2\pi]$, with initial condition $y_0 = 1$, obtained by applying the implicit Euler method with the variable-step-size strategy just described. The threshold values we use are $M = 10^{-6}$ and $N = 10^{-3}$ and an initial time step $k_0 = 10^{-2}$. In Figure 4.5 we plot the error $|v_n - y_n|$ of the solution.

Exercise 4.1 *Write a code that computes the solution to the example above and*
  (a) Recover the plots in figure 4.5.
  (b) Compute a table showing the times t_n at which the time step is adjusted and the corresponding new values of k_n.

Exercise 4.2 *The* trapezoidal rule *(or method) to approximate the ordinary initial value problem*

$$\frac{dy}{dt} = f(y, t),$$
$$y(0) = y_0,$$

is the implicit method given by:

$$\frac{v_{n+1} - v_n}{k} = \frac{f(v_{n+1}, t_{n+1}) + f(v_n, t_n)}{2}, \quad t_n = kn, \tag{4.17}$$
$$v_0 = y_0,$$

where, as usual, v_n denotes the approximation to $y(t_n)$.
  (a) Show that the stability region is $\{z \in \mathbb{C}, \ \mathrm{Re}\, z \leq 0\}$.
  (b) Show that (4.17) is a second-order accurate method.

CHAPTER 5

TWO-STEP AND MULTISTEP METHODS

The character of some equations, such as the equations of motion in a many-body problem in Newtonian mechanics, requires methods of high and very high orders of accuracy. The use of high-order one-step methods (such as Runge-Kutta methods) for such problems turns out to be computationally very expensive. For this reason multistep methods have been developed. These methods can be both of very high accuracy and fast. In this chapter we study the leapfrog method, which is the simplest of all, and Adams methods, which include both explicit and implicit methods. At the end we show a simple example of the predictor-corrector strategy for multistep methods.

5.1 Multistep methods

Consider the initial value problem

$$\frac{dy}{dt} = f(y, t),$$
$$y(0) = y_0.$$

(5.1)

Introduction to Numerical Methods for Time Dependent Differential Equations, First Edition. By Heinz-O. Kreiss and Omar E. Ortiz. Copyright © 2014 John Wiley & Sons, Inc.

Thus far, we have approximated equation (5.1) by one-step methods (see Definition 2.9). They have the form

$$v_{n+1} = v_n + k\Phi(v_n, t_n, k_n). \tag{5.2}$$

If we know v_n, we can calculate v_{n+1}. Another class of methods are the *multistep methods.*[1]

Definition 5.1 *A multistep method to approximate equation (5.1) is of the form*

$$v_{n+1} = \sum_{j=0}^{r} \alpha_j v_{n-j} + k \sum_{j=-1}^{r} \beta_j f(v_{n-j}, t_{n-j}), \tag{5.3}$$

where α_j and β_j are given constants and $r \geq 1$. The method is said to be explicit if $\beta_{-1} = 0$ and implicit otherwise.

The constants α_j and β_j, which determine a particular method, are chosen to have good accuracy and stability properties.

Notice that to apply a multistep method to the initial value problem (5.1) we set $v_0 = y_0$ and then need to prescribe the first r steps $v_1, v_2, \ldots, v_r$ by some supplementary rule, since the iteration needs the previous $r + 1$ values of v_j. This is a drawback of multistep methods compared to one-step methods if during calculation one has to change the step size to maintain the accuracy. In the latter all that is necessary to start the iteration is the initial condition $v_0 = y_0$.

Multistep methods are, however, preferred in applications where a very high degree of accuracy is necessary, such as for orbit calculation in many-body problems. The reason is the following. A one-step method would require several evaluations of the function $f(v, t)$ at every time step (e.g., Runge-Kutta methods) or, even worse, evaluations of the function and some of its derivatives (e.g., Taylor methods). On the other hand, a multistep method would require only one evaluation of the function $f(v, t)$ at every time step, because it reuses the evaluations performed in the preceding $r + 1$ steps. In many applications the evaluation of $f(v, t)$ is by far the most expensive part of the computation; then multistep methods are more efficient.

5.2 Leapfrog method

In this section we consider one particular method, the *leapfrog method,* which is given by

$$\frac{v_{n+1} - v_{n-1}}{2k} = f(v_n, t_n). \tag{5.4}$$

This method is the explicit two-step method that follows, with $r = 1$, from (5.3) with $\alpha_0 = 0$, $\alpha_1 = 1$, $\beta_{-1} = 0$, $\beta_0 = 2$, and $\beta_1 = 0$, that is, it can be written

$$v_{n+1} = v_{n-1} + 2kf(v_n, t_n). \tag{5.5}$$

---

[1]Methods of the form (5.3) are sometimes called *linear* multistep methods, because they use only linear combinations of the values v_j and $f(v_j, t_j)$.

It turns out that this method is very efficient for oscillator problems, such as

$$y_t = aiy + F(t), \quad a \in \mathbb{R}, \tag{5.6}$$

and also for systems of equations arising from hyperbolic partial differential equations.

The leapfrog method is an example of a two-step method: To compute v_{n+1}, one needs the previous two values, v_n and v_{n-1}. In particular, to start the computations one needs v_0 and v_1. Clearly, if an initial value condition $y(0) = y_0$ is given for (5.2), one can choose $v_0 = y_0$. To get v_1 one can use the explicit Euler method, for example. Then the starting data for (5.5) are

$$\begin{aligned} v_0 &= y_0, \\ v_1 &= v_0 + kf(v_0, 0) = y_0 + kf(y_0, 0). \end{aligned} \tag{5.7}$$

Exercise 5.1 *Why is (5.5), (5.7) second-order accurate?*

Truncation error. We will now calculate the truncation error R_n of the leapfrog method. By definition, r_n is obtained if one substitutes the exact solution $y(t)$ into the difference equations (5.4). Note that

$$y_{n+1} = y_n + k\left(\frac{dy}{dt}\right)_n + \frac{k^2}{2}\left(\frac{d^2y}{dt^2}\right)_n + \frac{k^3}{6}\left(\frac{d^3y}{dt^3}\right)_n + \frac{k^4}{24}\left(\frac{d^4y}{dt^4}\right)_n + \mathcal{O}(k^5),$$

$$y_{n-1} = y_n - k\left(\frac{dy}{dt}\right)_n + \frac{k^2}{2}\left(\frac{d^2y}{dt^2}\right)_n - \frac{k^3}{6}\left(\frac{d^3y}{dt^3}\right)_n + \frac{k^4}{24}\left(\frac{d^4y}{dt^4}\right)_n + \mathcal{O}(k^5).$$

Therefore,

$$\begin{aligned} R_n &= \frac{y_{n+1} - y_{n-1}}{2k} - f(y_n, t_n) \\ &= \left(\frac{dy}{dt}\right)_n - f(y_n, t_n) + \frac{k^2}{6}\left(\frac{d^3y}{dt^3}\right)_n + \mathcal{O}(k^4) \\ &= \frac{k^2}{6}\left(\frac{d^3y}{dt^3}\right)_n + \mathcal{O}(k^4). \end{aligned}$$

That is, to leading order the truncation error is $R_n \simeq (k^2/6)(d^3y/dt^3)_n$.

Stability. To calculate the stability of the leapfrog method, we apply it to the model equation,

$$\frac{du}{dt} = \lambda u.$$

The leapfrog approximation is

$$v_{n+1} = v_{n-1} + 2k\lambda v_n. \tag{5.8}$$

Clearly, the solution of (5.8) is determined uniquely by an initial condition

$$v_0 = \bar{v}_0, \quad v_1 = \bar{v}_1. \tag{5.9}$$

Generalizing Definition 3.6, we define the stability region as follows.

Definition 5.2 *The stability region of the leapfrog method consists of all complex numbers λk for which all solutions of* (5.8), (5.9) *are uniformly bounded for all* $n = 0, 1, 2, 3, \ldots$.

Solutions of difference equations. We first try to find a solution of (5.8) which is of the form

$$v_n = \kappa^n, \quad \kappa \neq 0. \tag{5.10}$$

Introducing (5.10) into (5.8) gives us

$$\kappa^{n+1} = \kappa^{n-1} + 2\lambda k \kappa^n.$$

Thus, (5.10) is a solution if and only if κ satisfies the so-called *characteristic equation*

$$\kappa^2 - 2\lambda k \kappa - 1 = 0. \tag{5.11}$$

Clearly,

$$\kappa_1 = \lambda k + \sqrt{1 + (\lambda k)^2} \quad \text{and} \quad \kappa_2 = \lambda k - \sqrt{1 + (\lambda k)^2}$$

are solutions of (5.11). Concerning the difference equation (5.5) with initial conditions (5.9), we have to distinguish between two cases.

1. $(\lambda k)^2 \neq -1$ (i.e., $\lambda k = \pm i$). In this case, $\kappa_1 \neq \kappa_2$, and the general solution of (5.8) is

$$v_n = \sigma_1 \kappa_1^n + \sigma_2 \kappa^2, \quad n = 0, 1, 2, \ldots. \tag{5.12}$$

   The constants σ_1, σ_2 are determined (5.9) by

$$\bar{v}_0 = \sigma_1 + \sigma_2, \quad \bar{v}_1 = \sigma_1 \kappa_1 + \sigma_2 \kappa_2.$$

2. $\lambda k = \pm i$. In this case there is only one solution to the characteristic equation, $\kappa_1 = \kappa_2 = \kappa = \lambda k$, which takes the value either $+i$ or $-i$. We get a one-parameter family of solutions $v_n = \sigma_1 \kappa^n$. In general, the two initial conditions (5.9) cannot be satisfied by $v_n = \sigma_1 \kappa^n$. We need to find a second solution. We try to find a solution of the form

$$v_n = n\kappa^n, \quad n = 0, 1, 2, \ldots. \tag{5.13}$$

Introducing (5.13) into (5.8) gives us

$$(n+1)\kappa^{n+1} = (n-1)\kappa^{n-1} + 2\lambda k n \kappa^n,$$

that is,

$$(n+1)\kappa^2 - 2n\lambda k \kappa - (n-1) = 0.$$

Since κ satisfies the characteristic equation (5.11), the last equation reduces to

$$\kappa^2 + 1 = 0. \tag{5.14}$$

By assumption we have $\lambda k = \pm i$; thus, $\kappa = \pm i$, and (5.14) is satisfied. Thus, for $\lambda k = \pm i$, the general solution of (5.8) is

$$v_n = \sigma_1 \kappa^n + \sigma_2 n \kappa^n \quad \text{with} \quad \kappa = \lambda k. \tag{5.15}$$

Again, we can determine σ_1, σ_2 so that the initial conditions (5.9) are fulfilled,

$$\bar{v}_0 = \sigma_1, \quad \bar{v}_1 = \sigma_1 \kappa + \sigma_2 \kappa.$$

Stability region. Since we have computed the general solution of the difference equation (5.8), we can now determine the stability region of the leapfrog method.

Theorem 5.3 *The stability region of the leapfrog method consists of all λk with*

$$\text{Re } \lambda k = 0 \quad \text{and} \quad |\text{Im } \lambda k| < 1.$$

Proof: First note that the two points $\lambda k = \pm i$ do not belong to the stability region because, by (5.15), the solution of (5.8) is generally unbounded. Now, let $\lambda k = ai$, $a \in \mathbb{R}$, $|a| < 1$. The roots of the characteristic equation (5.11) are

$$\kappa_{1,2} = ai \pm \sqrt{1 - a^2}$$

with $|\kappa_j|^2 = a^2 + (1 - a^2) = 1$. It follows that all solutions of (5.12) remain bounded and the method is stable.

Conversely, assume that λk belongs to the stability region. For the roots $\kappa_{1,2}$ of the characteristic equation we have

$$\kappa_1, \kappa_2 = -1, \quad \kappa_1 + \kappa_2 = 2\lambda k.$$

Since $|\kappa_1| \le 1$ and $|\kappa_2| \le 1$ we find that $|\kappa_1| = |\kappa_2|^{-1}$ and then $|\kappa_1| = |\kappa_2| = 1$. Thus,

$$\kappa_1 = e^{i\alpha}, \quad \kappa_2 = -\frac{1}{\kappa_1} = -e^{-i\alpha}, \quad \alpha \in \mathbb{R}.$$

Therefore,

$$2\lambda k = \kappa_1 + \kappa_2 = 2i \sin(\alpha),$$

and the assertion $\lambda k = ia$, $a \in \mathbb{R}$, $|a| < 1$ follows. ∎

Modified leapfrog method. Since the stability region for the leapfrog method lies on the imaginary axis, one can expect the method to work well for oscillatory problems such as (5.6). For other problems we have to modify it. Consider, for example,

$$\frac{dy}{dt} = \lambda y + F(t), \quad \lambda = \eta + i\xi, \quad \eta, \xi \in \mathbb{R}, \; \eta \le 0. \tag{5.16}$$

One can approximate (5.16) by

$$\frac{v_{n+1} - v_{n-1}}{2k} = \lambda v_n + F_n + \frac{\eta}{2}(v_{n+1} - 2v_n + v_{n-1}),$$

which is equivalent to

$$(1 - k\eta)v_{n+1} = (1 + k\eta)v_{n-1} + 2ki\xi v_n + 2F_n. \tag{5.17}$$

The modified method is second-order accurate.

Exercise 5.2 *Prove that the stability region of the modified leapfrog method* (5.16) *includes the semiunit disk* $x^2 + y^2 < 1$, $x \le 0$.

5.3 Adams methods

Definition 5.4 *The* Adams methods *are multistep methods with* $\alpha_0 = 1$ *and* $\alpha_j = 0$ *for* $j = 1, 2, 3, \ldots, r$. *They are of the form*

$$v_{n+1} = v_n + k \sum_{j=-1}^{r} \beta_j f(v_{n-j}, t_{n-j}). \tag{5.18}$$

The constants β_j are chosen to get the accuracy and stability properties desired. For example, the *Adams-Bashforth* are explicit Adams methods; that is, they choose $\beta_{-1} = 0$, and β_j with $j = 0, 1, \ldots, r$ so that the accuracy order of the method is as high as possible. It can be shown that an equivalent way to obtain the coefficients is as follows. Integrating equation (5.1) between t_n and t_{n+1} gives

$$y_{n+1} = y_n + \int_{t_n}^{t_{n+1}} f(y, t) \, dt.$$

If we replace $f(y, t)$ in the integral by its interpolating polynomial of degree r, at the grid points t_n, t_{n-1}, $t_{n-2}, \ldots, t_{n-r}$, replace y by v as usual, and integrate, we get the Adams-Bashforth method.

Exercise 5.3 *Show that the one-step Adams-Bashforth method is just the explicit Euler method and derive the two-step Adams-Bashforth method.* Hint: *Instead of using the general equation* (5.1), *use the linear equation* $dy/dt = \lambda y$.

We derive as an example the four-step Adams-Bashforth method, (i.e. with $r = 3$). It is enough to consider the linear equation $dy/dt = \lambda y$. Inserting the exact solution into (5.18) and using Taylor expansions centered in y_n we see that we can cancel the first four terms of the local error expansion if the following equations are satisfied

$$\begin{aligned}
\beta_0 + \beta_1 + \beta_2 + \beta_3 &= 1, \\
-\beta_1 - 2\beta_2 - 3\beta_3 &= \tfrac{1}{2}, \\
\tfrac{1}{2}\beta_1 + 2\beta_2 + \tfrac{9}{2}\beta_3 &= \tfrac{1}{6}, \\
-\tfrac{1}{6}\beta_1 - \tfrac{4}{3}\beta_2 - \tfrac{9}{2}\beta_3 &= \tfrac{1}{24}.
\end{aligned} \tag{5.19}$$

These equations have the unique solution $\beta_0 = 55/24$, $\beta_1 = -59/24$, $\beta_2 = 37/24$, and $\beta_3 = -3/8$, which gives the fourth-order accurate method

$$v_{n+1} = v_n + \frac{k}{24}\Big(55f(v_n, t_n) - 59f(v_{n-1}, t_{n-1})$$
$$+ 37f(v_{n-2}, t_{n-2}) - 9f(v_{n-3}, t_{n-3})\Big). \quad (5.20)$$

In general, the β_j, $j = 0, 1, \ldots, r$, coefficients in a $(r + 1)$-step Adams-Bashforth method can be chosen to cancel terms in the local error expansion up to $\mathcal{O}(k^{r+1})$. The resulting method is accurate of order $r + 1$.

If one chooses $\beta_{-1} \neq 0$ in (5.18), on obtains implicit methods called *Adams-Moulton methods*. In this case one has $r + 2$ constants to determine and one further term of the local error expansion can be canceled (compared to the Adams-Bashforth methods), thus obtaining an $(r + 2)$-order accurate method. For example, with a fourth-step Adams-Moulton method we can cancel the first five terms in the local error expansion. The system of equations to satisfy is

$$\beta_{-1} + \beta_0 + \beta_1 + \beta_2 + \beta_3 = 1,$$
$$\beta_{-1} - \beta_1 - 2\beta_2 - 3\beta_3 = \tfrac{1}{2},$$
$$\tfrac{1}{2}\beta_{-1} + \tfrac{1}{2}\beta_1 + 2\beta_2 + \tfrac{9}{2}\beta_3 = \tfrac{1}{6}, \quad (5.21)$$
$$\tfrac{1}{6}\beta_{-1} - \tfrac{1}{6}\beta_1 - \tfrac{4}{3}\beta_2 - \tfrac{9}{2}\beta_3 = \tfrac{1}{24},$$
$$\tfrac{1}{24}\beta_{-1} + \tfrac{1}{24}\beta_1 + \tfrac{2}{3}\beta_2 + \tfrac{27}{8}\beta_3 = \tfrac{1}{120},$$

whose unique solution is $\beta_{-1} = 251/720$, $\beta_0 = 323/360$, $\beta_1 = -11/30$, $\beta_2 = 53/360$, and $\beta_3 = -19/720$. The method is fifth-order accurate, given by

$$v_{n+1} = v_n + \frac{k}{720}\Big(251f(v_{n+1}, t_{n+1}) + 646f(v_n, t_n) - 264f(v_{n-1}, t_{n-1})$$
$$+ 106f(v_{n-2}, t_{n-2}) - 19f(v_{n-3}, t_{n-3})\Big). \quad (5.22)$$

Exercise 5.4 *Derive the systems of equations* (5.19) *and* (5.21).

Initial steps. To apply a multistep method one needs to obtain the first $r - 1$ steps with a different method. A possibility is to use a one-step method of the same order. For example, to apply (5.20), one could compute v_1, v_2, and v_3, applying three steps of the classical fourth-order Runge-Kutta method (3.5).

5.4   Stability of multistep methods

Consider the model problem

$$\frac{dy}{dt} = \lambda y,$$
$$y(0) = y_0. \quad (5.23)$$

Definition 5.5 *Multistep method* (5.3) *is said to be stable if when applied to problem* (5.4), *the solution* v_n *stays bounded for all* $n = 0, 1, 2, \ldots$.

One can proceed as in the leapfrog method to find solutions of the multistep method. One first looks for solutions of the form $v_n = \kappa^n$. Plugging v_n into the difference equations (5.3) for the particular problem (5.23) gives

$$p_\mu(\kappa) = (1 - \beta_{-1}\mu)\kappa^{r+1} - \sum_{j=0}^{r}(\alpha_j + \beta_j\mu)\kappa^{r-j} = 0, \quad \mu = \lambda k. \quad (5.24)$$

This polynomial equation of degree $r+1$ in κ, whose coefficients are linear functions of μ, is sometimes called a *characteristic equation*. When $p_\mu(\kappa)$ has $r + 1$ distinct roots $\kappa_j, \; j = 1, 2, \ldots, r + 1$, the general solution of the multi-step method can be written as

$$v_n = \sum_{j=1}^{r+1} c_j \kappa_j^n.$$

In this case the method is clearly stable when $|\kappa_j| \leq 1, \; j = 1, 2, \ldots, r + 1$. When some of the roots κ_j are multiple roots (i.e., have multiplicity greater than 1), some other independent solutions of the form

$$n^l \kappa_j^n, \quad l \in \mathbb{Z}, \quad l \leq \text{multiplicity of } \kappa_j,$$

are needed to expand the general solution v_n. The general solution will be bounded provided that these multiple roots of (5.24) satisfy $|\kappa_j| < 1$. Then we have

Lemma 5.6 *The multistep method* (5.3) *is stable if the solutions* κ_j *of* (5.24) *satisfy*

$$\begin{aligned} |\kappa_j| \leq 1 & \quad \text{for single roots of } p_\mu(\kappa), \\ |\kappa_j| < 1 & \quad \text{for multiple roots of } p_\mu(\kappa). \end{aligned} \quad (5.25)$$

The *stability region* of the multistep method consists of all the μ in the complex plane such that the method is stable. The fact that all the coefficients of $p_\mu(\kappa)$ are linear functions of μ allows one to apply a simple procedure, known as the *boundary locus method*, to find the stability region. To understand this method, notice first that $p_\mu(\kappa) = 0$ can be written as

$$p_\alpha(\kappa) = \mu p_\beta(\kappa), \quad (5.26)$$

where

$$p_\alpha(\kappa) = \kappa^{r+1} - \sum_{j=0}^{r} \alpha_j \kappa^{r-j}, \quad p_\beta(\kappa) = \sum_{j=-1}^{r} \beta_j \kappa^{r-j}.$$

By continuity, because of (5.25), for every μ on the boundary of the stability region, at least one of the eigenvalues κ_j satisfies $|\kappa_j| = 1$. Writing this eigenvalue as $\kappa = e^{i\theta}$ and using (5.26), we have, for the μ on the boundary,

$$\mu = \frac{p_\alpha(e^{i\theta})}{p_\beta(e^{i\theta})}. \quad (5.27)$$

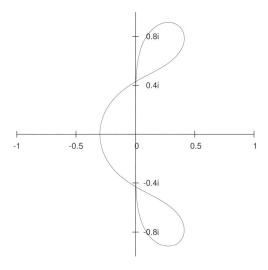

Figure 5.1   Boundary locus for the four-step Adams-Bashforth method.

Therefore, to find the boundary of the stability region, one can plot the curve (5.27) for $\theta \in [0, 2\pi]$. This is clearly a closed curve that splits the complex μ-plane in two or more domains. Then, by computing all the roots of $p_\mu(\kappa)$ in one arbitrary interior point on every domain to check the condition (5.25), one can identify which of these domains belong to the stability region and which of them do not. Consider, for example, the four-step Adams-Bashforth method (5.20); we have

$$p_\alpha(e^{i\theta}) = e^{i4\theta} - e^{i3\theta}, \qquad p_\beta(e^{i\theta}) = \frac{55}{24}e^{i3\theta} - \frac{59}{24}e^{i2\theta} + \frac{37}{24}e^{i\theta} - \frac{9}{24}.$$

The curve (5.27) divides the complex plane into four domains (see Figure 5.1).

Exercise 5.5 *Show that the only domain of Figure 5.1 belonging to the stability region of the four-step Adams-Bashforth method is the one just left of the origin.*

Predictor-corrector strategy.   In many applications it might be interesting to use implicit multistep methods such as the Adams-Moulton methods introduced earlier. However, the implicit character of these methods implies that one needs to solve large systems of algebraic equations, which become complicated or computationally expensive. A procedure designed to overcome this difficulty is to use an implicit method in combination with an explicit method as a predictor. Here we simply give an example that uses a third-order Adams-Bashforth method as a predictor and a fourth-order Adams-Moulton method as a corrector. If one wants to solve the equation

$$\frac{dy}{dt} = f(y, t),$$

this predictor-corrector scheme is

$$
\tilde{v}_{n+1} = v_n + \frac{k}{12}\Big(23f(v_n, t_n) - 16f(v_{n-1}, t_{n-1})
$$
$$
+ 5f(v_{n-2}, t_{n-2})\Big), \qquad\qquad \text{(predictor)}
$$
$$
v_{n+1} = \frac{k}{24}\Big(9f(\tilde{v}_{n+1}, t_{n+1}) + 19f(v_n, t_n)
$$
$$
- 5f(v_{n-1}, t_{n-1}) + f(v_{n-2}, t_{n-2})\Big). \qquad \text{(corrector)}
$$

The resulting scheme is clearly explicit.

CHAPTER 6

SYSTEMS OF DIFFERENTIAL EQUATIONS

In this chapter we motivate, by means of a simple observation, the generalization to systems of equations of what we have learned for scalar differential equations.

Consider the initial value problem for a system of differential equations

$$\frac{d\mathbf{y}}{dt} = \mathbf{f}(\mathbf{y}, t),$$

$$\mathbf{y}(0) = \mathbf{y}_0,$$

$$(6.1)$$

where $\mathbf{y}$ is an N-component vector and $\mathbf{f}$ is an N-component vector-valued function:

$$\mathbf{y} = \begin{pmatrix} y^1 \\ \vdots \\ y^N \end{pmatrix} \quad \text{and} \quad \mathbf{f}(\mathbf{y}, t) = \begin{pmatrix} f^1(\mathbf{y}, t) \\ \vdots \\ f^N(\mathbf{y}, t) \end{pmatrix}.$$

The simplest systems are linear systems with constant coefficients:

$$\frac{d\mathbf{y}}{dt} = A\mathbf{y} + \mathbf{F}(t), \quad \mathbf{y}(0) = \mathbf{y}_0. \tag{6.2}$$

Here A is a constant $N \times N$-matrix and $\mathbf{F}$ has N components.

Introduction to Numerical Methods for Time Dependent Differential Equations, First Edition.
By Heinz-O. Kreiss and Omar E. Ortiz. Copyright © 2014 John Wiley & Sons, Inc.

Let us assume that A has a complete set of eigenvectors; that is, there exists a transformation S that transforms A into a diagonal matrix Λ:

$$\Lambda = S^{-1}AS, \qquad \Lambda = \begin{pmatrix} \lambda_1 & 0 & 0 & \ldots & 0 \\ 0 & \lambda_2 & 0 & \ldots & 0 \\ & & \ddots & \ddots & \ddots \\ 0 & & \ldots\ldots & 0 & \lambda_N \end{pmatrix}.$$

Exercise 6.1 *The exponential of a matrix $A \in \mathbb{C}^{N \times N}$ is defined via the Taylor series,*

$$\exp(A) = \sum_{j=0}^{\infty} \frac{1}{j!} A^j.$$

(a) Show that $\exp(A)$ is well defined (i.e., the series is always convergent).
(b) Show that

$$\frac{d}{dt} \exp(At) = A \exp(At).$$

(c) Show that Lemma 1.2 generalizes for problem (6.2) in the form

$$\mathbf{y}(t) = \exp(At)\mathbf{y}_0 + \int_0^t \exp(A(t-s))\, \mathbf{F}(s)\, ds.$$

Introducing a new variable,

$$\tilde{\mathbf{y}} = S^{-1}\mathbf{y},$$

into (6.2), we obtain

$$\frac{d\tilde{\mathbf{y}}}{dt} = \Lambda \tilde{\mathbf{y}} + \tilde{\mathbf{F}}(t), \quad \tilde{\mathbf{y}}(0) = \tilde{\mathbf{y}}_0. \tag{6.3}$$

with $\tilde{\mathbf{F}}(t) = S^{-1}\mathbf{F}(t)$ and $\tilde{\mathbf{y}}_0 = S^{-1}\mathbf{y}_0$. Since Λ is diagonal, the system above reads in component form

$$\tilde{y}^j = \lambda_j \tilde{y}^j + \tilde{F}^j(t), \quad \tilde{y}^j(0) = \tilde{y}_0^j, \quad j = 1, 2, \ldots, N. \tag{6.4}$$

In other words, (6.2) is equivalent to N decoupled scalar equations.

Assume that we approximate every scalar equation in (6.4) with the explicit Euler method on the same grid (i.e., with the same step size k),

$$\tilde{v}_{n+1}^j = (1 + \lambda_j k)\tilde{v}_n^j + kF_n^j, \quad \tilde{v}_0^j = \tilde{y}_0^j.$$

These equations read, in vector form,

$$\tilde{\mathbf{v}}_{n+1} = (I + k\Lambda)\tilde{\mathbf{v}}_n + k\tilde{\mathbf{F}}_n, \quad \tilde{\mathbf{v}}_n = \begin{pmatrix} \tilde{v}_n^1 \\ \vdots \\ \tilde{v}_n^N \end{pmatrix}. \tag{6.5}$$

Then, by using the same transformation S, we have

$$\mathbf{v}_n = S\tilde{\mathbf{v}}_n, \quad \tilde{\mathbf{F}}_n = S\mathbf{F}_n,$$

and (6.5) becomes

$$\mathbf{v}_{n+1} = (I + kA)\mathbf{v}_n + \mathbf{F}_n, \quad \mathbf{v}_0 = \mathbf{y}_0, \tag{6.6}$$

which is precisely the definition of the explicit Euler method to approximate the vector system of equations (6.2).

A consequence of these observations is that our analysis of Euler's method (Chapter 2) for scalar equations carries over to the linear systems described above. The same remark applies to all other methods that we have discussed.

In particular, if $\operatorname{Re}\lambda_j < 0$ for all $j = 0, \ldots, N$, we choose the step size k so that $k\lambda_j$ belongs to the stability region of the method for every $j = 1, \ldots, N$. Under this restriction, the size of the error $e_n = y_n - v_n$ can be determined by a truncation error analysis similar to the scalar case.

Consider now the nonlinear problem (6.1) and assume that the vector function f is smooth in its arguments. To study how this system behaves, we linearize the problem around the initial data vector given. This is, we write the solution as

$$\mathbf{y}(t) = \mathbf{y}_0 + \mathbf{u}(t),$$

and use Taylor expansion for $\mathbf{f}$. We obtain

$$\frac{d\mathbf{u}}{dt} = \mathbf{f}(\mathbf{y}_0, t) + \frac{\partial \mathbf{f}}{\partial \mathbf{y}}(\mathbf{y}_0, t)\mathbf{u} + \mathcal{O}(|\mathbf{u}|^2).$$

We assume that the Jacobian matrix of $\mathbf{f}$,

$$A(t) = \frac{\partial \mathbf{f}}{\partial \mathbf{y}}(\mathbf{y}_0, t),$$

is not singular; then for short times the nonlinear problem is governed by the linear problem

$$\frac{d\mathbf{u}}{dt} = A(t)\mathbf{u} + \mathbf{F}(t), \quad \mathbf{u}(0) = 0, \tag{6.7}$$

where $\mathbf{F}(t) = \mathbf{f}(\mathbf{y}_0, t)$.

We remind the reader of the definition of a Euclidean vector norm and its subordinate or induced matrix norm:

Definition 6.1 *The Euclidean vector norm of* $\mathbf{u} \in \mathbb{C}^n$ *is*

$$|\mathbf{u}| = \sqrt{\sum_{i=1}^{n} |u_i|^2},$$

and the "subordinate" or "induced" matrix norm of $A \in \mathbb{C}^{n \times n}$ *is*

$$|A| = \max_{\mathbf{u}}\{|A\mathbf{u}|, \text{ s.t. } \mathbf{u} \in \mathbb{C}^n \text{ with } |\mathbf{u}| = 1\}.$$

Exercise 6.2 *Show that the vector norm and its subordinate matrix norm satisfy*

$$|A\mathbf{u}| \leq |A|\,|\mathbf{u}|.$$

We assume now that the matrix $A(t)$ has a complete system of eigenvectors which are uniformly independent in the time interval of interest, that is, there exists a smooth transformation matrix $S(t)$ such that

$$S^{-1}(t)AS(t) = \Lambda(t) = \mathrm{diag}(\lambda_1(t), \ldots, \lambda_N(t))$$

and

$$|S(t)| + |S^{-1}(t)| < \mathrm{const}. \tag{6.8}$$

If $\mathrm{Re}\,\lambda_j(t) < 0$, $j = 1, 2, \ldots, N$, during some time interval $0 \leq t \leq T$, the solution of (6.1) is driven during some time by N linear, stable, scalar equations. We can integrate numerically in time using any of the methods we studied for scalar ODEs provided that we choose the time step k small enough so that $k\lambda_j(t)$ belongs in the stability region for $j = 1, 2, \ldots, N$. From a computational point of view, an implicit assumption is that the norm of the transformation $S(t)$ and that of its inverse are not very large. In other words, the transformation should not be almost singular at any time.

In real computations, it is very important to carry out tests to check whether or not our numerical scheme converges in the time interval of interest. For example, we first compute a solution in a time interval using a time step k. Then we can use half that time step and check that we can compute the solution during the same time interval or at least that the interval of existence of our numerical solution does not go to zero as the time step diminishes. Of course, one can also perform the precision quotient tests of Section 2.6 using any vector norm instead of absolute values.

For completeness we write the explicit Euler method that approximates the general problem (6.1):

$$\mathbf{v}_{n+1} = \mathbf{v}_n + k\mathbf{f}(\mathbf{v}_n, t_n), \quad \mathbf{v}_0 = \mathbf{y}_0.$$

PART II

PARTIAL DIFFERENTIAL EQUATIONS AND THEIR APPROXIMATIONS

CHAPTER 7

FOURIER SERIES AND INTERPOLATION

Fourier series are among the most useful ideas in mathematics, for many reasons. For the problems we are interested in in this book, Fourier theory is essential to study the stability of numerical approximations and to introduce a family of high-precision methods, the spectral and pseudo-spectral methods. In this chapter we review briefly Fourier series and their ability to represent functions both exactly or as an approximation. We also study Fourier interpolation, which can be thought of as a version of Fourier series for discrete functions.

7.1 Fourier expansion

In this section we consider the expansion of 1-periodic function $f(x)$ into Fourier series. Here $f : \mathbb{R} \to \mathbb{C}$ is called 1-*periodic* if $f(x+1) = f(x)$ for all x. Important examples of 1-periodic functions are the exponentials,

$$e^{2\pi i \omega x} = \cos(2\pi \omega x) + i \sin(2\pi \omega x), \quad \omega = 0, \pm 1, \pm 2, \ldots,$$

Introduction to Numerical Methods for Time Dependent Differential Equations, First Edition. **81** By Heinz-O. Kreiss and Omar E. Ortiz. Copyright © 2014 John Wiley & Sons, Inc.

which play a central role in mathematics. One reason is that differentiation results in multiplication by a constant factor,

$$\frac{d}{dx}e^{2\pi i \omega x} = 2\pi i \omega e^{2\pi i \omega x}.$$

As a consequence, by the use of exponentials, one can often transform a differential equation into an algebraic equation.

For this reason it is important that quite general 1-periodic functions $f(x)$ can be written as *Fourier series*:

$$f(x) = \sum_{\omega=-\infty}^{\infty} e^{2\pi i \omega x} \hat{f}(\omega). \tag{7.1}$$

The series converges rapidly if $f(x)$ is smooth.

The complex numbers $\hat{f}(\omega)$ are uniquely determined by $f(x)$. Assuming the representation (7.1), this can be seen as follows. Let ω_1 be fixed. Then for all frequencies ω,

$$\int_0^1 e^{-2\pi i \omega_1 x} e^{2\pi i \omega x}\,dx = \begin{cases} 0 & \text{if } \omega \neq \omega_1 \\ 1 & \text{if } \omega = \omega_1. \end{cases} \tag{7.2}$$

Therefore, by (7.1),[1]

$$\int_0^1 e^{-2\pi i \omega_1 x} f(x)\,dx = \sum_{\omega=-\infty}^{\infty} \int_0^1 e^{-2\pi i \omega_1 x + 2\pi i \omega x}\,dx\,\hat{f}(\omega) = \hat{f}(\omega_1), \tag{7.3}$$

which holds for all frequencies. We call the discrete function $\hat{f}(\omega)$, $\omega = 0, \pm 1, \pm 2, \dots$ the Fourier coefficients of $f(x)$.

By (7.1), we can recover $f(x)$ if we know the Fourier coefficients of $f(x)$. The representation (7.1) is valid if $f(x)$ is only piecewise smooth. Jump discontinuities of $f(x)$ are allowed. However, if $f(x)$ is discontinuous, the convergence is slow.

In the following theorem we state, without proof, some basic results about convergence of Fourier series (see, e.g., [2, 10, 9]).

Definition 7.1 *We denote by $C^n[a, b]$ the space of functions that are n times, $n \geq 1$, continuously differentiable in the interval $a \leq x \leq b$; and similarly, $C^n(a, b)$ when the interval is $a < x < b$.*

Theorem 7.2 *Let $f(x) \in C^1(-\infty, \infty)$ be 1-periodic. Then the Fourier expansion (7.1) holds and converges uniformly to $f(x)$. If $f(x)$ is only piecewise continuously differentiable, (7.1) converges uniformly in any subinterval $[a, b]$ if $f(x) \in C^1[a, b]$, while at the points of discontinuity, the Fourier series converges to $\frac{1}{2}\big(f(x+0) + f(x-0)\big)$.*

---

[1] The interchange of summation and integration can be justified under quite general assumptions on f. It suffices to assume that $\int_0^1 |f(x)|^2\,dx$ is finite.

The decay of the Fourier coefficients, as $|\omega| \to \infty$, is related to the smoothness of the function; the following theorem is a general result about this.

Lemma 7.3 *If $f(x) \in C^n(-\infty, \infty)$ is 1-periodic, then*

$$|\hat{f}(0)| \leq \int_0^1 |f(x)| \, dx$$

and

$$|\hat{f}(\omega)| \leq \frac{1}{(2\pi|\omega|)^n} \int_0^1 \left|\frac{d^n f}{dx^n}\right| \, dx, \quad \text{for } \omega \neq 0.$$

Proof: For $\omega = 0$ we have

$$|\hat{f}(0)| = \left|\int_0^1 f(x) \, dx\right| \leq \int_0^1 |f(x)| \, dx.$$

For $\omega \neq 0$ we integrate by parts and note that the contribution from the boundary terms is zero because of the periodicity of f:

$$
\begin{aligned}
\hat{f}(\omega) &= \int_0^1 e^{-2\pi i \omega x} f(x) \, dx \\
&= -\frac{1}{2\pi i \omega} e^{-2\pi i \omega x} f(x) \Big|_0^1 + \int_0^1 \frac{e^{-2\pi i \omega x}}{2\pi i \omega} \frac{df}{dx} \, dx \\
&= \int_0^1 \frac{e^{-2\pi i \omega x}}{2\pi i \omega} \frac{df}{dx} \, dx \\
&= -\frac{1}{(2\pi i \omega)^2} e^{-2\pi i \omega x} \frac{df}{dx} \Big|_0^1 + \int_0^1 \frac{e^{-2\pi i \omega x}}{(2\pi i \omega)^2} \frac{d^2 f}{dx^2} \, dx \\
&= \int_0^1 \frac{e^{-2\pi i \omega x}}{(2\pi i \omega)^2} \frac{d^2 f}{dx^2} \, dx \\
&= \int_0^1 \frac{e^{-2\pi i \omega x}}{(2\pi i \omega)^n} \frac{d^n f}{dx^n} \, dx,
\end{aligned}
$$

that is,

$$|\hat{f}(\omega)| \leq \left|\int_0^1 \frac{e^{-2\pi i \omega x}}{(2\pi \omega)^n} \frac{d^n f}{dx^n} \, dx\right| \leq \frac{1}{(2\pi|\omega|)^n} \int_0^1 \left|\frac{d^n f}{dx^n}\right| \, dx. \qquad \blacksquare$$

It is sometimes convenient to work with real quantities. If $f(x)$ is a real function, we have

$$f(x) = \sum_{\omega=-\infty}^{\infty} e^{2\pi i \omega x} \hat{f}(\omega)$$

$$= \bar{f}(x)$$

$$= \sum_{\omega=-\infty}^{\infty} e^{-2\pi i \omega x} \bar{\hat{f}}(\omega)$$

$$= \sum_{\omega=-\infty}^{\infty} e^{2\pi i \omega x} \bar{\hat{f}}(-\omega),$$

so that the uniqueness of the Fourier coefficients implies that

$$\hat{f}(\omega) = \bar{\hat{f}}(-\omega)$$

or, equivalently,

$$\bar{\hat{f}}(\omega) = \hat{f}(-\omega).$$

Therefore,

$$f(x) = \hat{f}(0) + \sum_{\omega=1}^{\infty} \left(e^{2\pi i \omega x} \hat{f}(\omega) + e^{-2\pi i \omega x} \hat{f}(-\omega) \right)$$

$$= \hat{f}(0) + \sum_{\omega=1}^{\infty} \left(\bar{\hat{f}}(\omega) + \hat{f}(\omega) \right) \cos(2\pi \omega x) + i \left(\hat{f}(\omega) - \bar{\hat{f}}(\omega) \right) \sin(2\pi \omega x).$$

The numbers

$$a(\omega) = \bar{\hat{f}}(\omega) + \hat{f}(\omega) = 2 \int_0^1 \cos(2\pi \omega x) f(x) \, dx,$$

$$b(\omega) = i \left(\hat{f}(\omega) - \bar{\hat{f}}(\omega) \right) = 2 \int_0^1 \sin(2\pi \omega x) f(x) \, dx,$$

$$\hat{f}(0) = \int_0^1 f(x) \, dx = a(0)$$

are real. We have thus obtained an expansion of $f(x)$ in terms of $\cos(2\pi \omega x)$ and $\sin(2\pi \omega x)$ with real coefficients.

We illustrate Fourier series with some examples. Consider the piecewise constant, 1-periodic discontinuous function

$$g(x) = \begin{cases} 1 & \text{for } 0 < x < \frac{1}{2}, \\ 0 & \text{for } x = 0, \text{ and } x = \frac{1}{2}, \\ -1 & \text{for } \frac{1}{2} < x < 1, \end{cases} \qquad g(x+1) = g(x). \qquad (7.4)$$

For $\nu = 0$, we obtain

$$\hat{g}(0) = \int_0^1 g(x)dx = 0.$$

For $\nu \neq 0$, we have

$$\hat{g}(\nu) = \int_0^1 e^{-2\pi i \nu x} g(x)\, dx$$

$$= \int_0^{1/2} e^{-2\pi i \nu x}\, dx - \int_{1/2}^1 e^{-2\pi i \nu x}\, dx$$

$$= -\frac{1}{2\pi i \nu} e^{-2\pi i \nu x}\big|_0^{1/2} + \frac{1}{2\pi i \nu} e^{-2\pi i \nu x}\big|_{1/2}^1$$

$$= \frac{1}{2\pi i \nu}(1 - e^{-\pi i \nu}) + \frac{1}{2\pi i \nu}(1 - e^{-\pi i \nu})$$

$$= \frac{1}{\pi i \nu}(1 - (-1)^\nu).$$

Thus,

$$g(x) = \sum_{\omega=-\infty}^{\infty} \hat{g}(\omega) e^{2\pi i \omega x}$$

$$= \sum_{\omega=1}^{\infty} \left(\hat{g}(\omega) e^{2\pi i \omega x} + \hat{g}(-\omega) e^{-2\pi i \omega x} \right)$$

$$= \sum_{\omega=1}^{\infty} \frac{1}{\pi i \omega}(1 - (-1)^\omega) \left(e^{2\pi i \omega x} - e^{-2\pi i \omega x} \right)$$

$$= \sum_{\omega=1}^{\infty} \frac{2(1 - (-1)^\omega)}{\pi \omega} \sin(2\pi \omega x)$$

$$= \frac{4}{\pi} \sum_{\omega=0}^{\infty} \frac{1}{2\omega + 1} \sin\big(2\pi(2\omega + 1)x\big).$$

The terms of the series decay only as $(2\omega + 1)^{-1}$ for $\omega \to \infty$. Therefore, the convergence of the series is slow.

In Figure 7.1 we plot the approximations of $g(x)$ by the truncated Fourier series

$$g^N(x) = \frac{4}{\pi} \sum_{\omega=0}^{N} \frac{1}{2\omega + 1} \sin\big(2\pi(2\omega + 1)x\big) \tag{7.5}$$

for $N = 10$ and $N = 100$. It can be seen in Figure 7.1 that on each side of the jump discontinuity of $g(x)$, the truncated Fourier series $g^N(x)$ shows an overshoot or peak that deviates more from the function than it does far from the discontinuity. Moreover, the height of the peak does not seem to go to zero when N increases. In fact, one can show that this peak remains finite in height even in the limit $N \to$

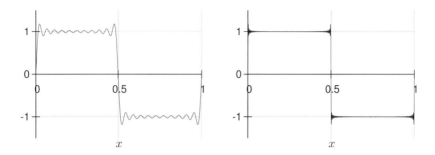

Figure 7.1 $g^{10}(x)$ on the left and $g^{100}(x)$ on the right.

∞. This behavior of Fourier series at jump discontinuities is known as the *Gibbs phenomenon*.

Exercise 7.1 *In this exercise we want to compute the Gibbs phenomenon for the previous example. To this end, compute the height of the first peak (i.e., right to zero and closest to the origin) by finding the maximum of the truncated Fourier series (7.5); you should obtain*

$$x_m = \frac{1}{4(N+1)} \quad and \quad g^N(x_m) = \frac{2}{\pi} \sum_{\omega=0}^{N} \frac{\sin\left((\pi/(N+1))(\omega + \frac{1}{2})\right)}{(\omega + \frac{1}{2})}.$$

Notice that the previous expression can be seen as the midpoint-rule approximation to the integral

$$\frac{2}{\pi} \int_0^\pi \frac{\sin(x)}{x} \, dx = 1 + \delta \quad with \ \delta \simeq 0.179$$

when the interval $[0, \pi]$ is divided into $N+1$ equal subintervals. In the limit $N \to \infty$ the overshoot becomes precisely $\delta > 0$.

If we integrate $g(x)$, we obtain a continuous function

$$h(x) = \int_0^x g(s) \, ds = \begin{cases} x & \text{for } 0 \le x \le \frac{1}{2} \\ 1 - x & \text{for } \frac{1}{2} \le x \le 1, \end{cases} \quad h(x) = h(x+1)$$

(see Figure 7.2). Integrating the terms of the Fourier series of $y(x)$ gives us

$$h(x) = \frac{4}{\pi} \sum_{\omega=0}^{\infty} \frac{1}{2\omega + 1} \int_0^x \sin\left(2\pi(2\omega + 1)s\right) ds$$

$$= \frac{4}{2\pi^2} \sum_{\omega=0}^{\infty} \frac{1}{(2\omega + 1)^2} \left(1 - \cos\left(2\pi(2\omega + 1)x\right)\right)$$

$$= \frac{1}{4} - \frac{2}{\pi^2} \sum_{\omega=0}^{\infty} \frac{1}{(2\omega + 1)^2} \cos\left(2\pi(2\omega + 1)x\right)$$

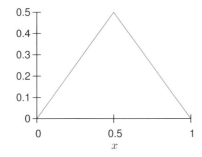

Figure 7.2   Continuous (piecewise smooth) function.

since the constant term is $\int_0^1 h(x)\,dx = \frac{1}{4}$. We see that the terms in the series decay like $(1+2\omega)^{-2}$ as $\omega \to \infty$ [i.e. the decay is faster than the series for $g(x)$].

In Figure 7.3 we show the truncated Fourier series approximating $h(x)$,

$$h^N(x) = \frac{1}{4} - \frac{2}{\pi^2} \sum_{\omega=0}^{N} \frac{1}{(2\omega+1)^2} \cos\bigl(2\pi(2\omega+1)x\bigr)$$

for $N = 1$ and $N = 10$. The function $h(x)$ is smoother than $g(x)$ and therefore the Fourier series converges faster.

7.2 L_2-norm and scalar product

In the first part of this book we have discussed the fundamental concepts that govern the theory of ordinary differential equations and their numerical solution. The results could often be expressed in terms of estimates in the maximum norm. For time-dependent partial differential equations, in particular for problems of wave propaga-

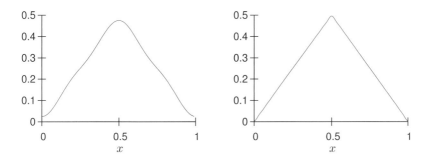

Figure 7.3 $h^1(x)$ on the left and $h^{10}(x)$ on the right.

tion, it is much more powerful to express the results in terms of the L_2-norm and its scalar product. The reason is that they are often closely related to estimates and concepts of the underlying physics.

They are also connected closely with Fourier expansions which we obtain using the concept of frozen coefficients. In this section we discuss this connection. Let $f(x) \in C^\infty(-\infty, \infty)$, $g(x) \in C^\infty(-\infty, \infty)$ be complex-valued 1-periodic functions. We define the L_2-scalar product and norm by

$$(f, g) = \int_0^1 \bar{f}(x) g(x) \, dx, \quad \|f\|^2 = (f, f), \quad \|f\| \geq 0. \tag{7.6}$$

The scalar product is a bilinear form, that is,

$$\begin{aligned}
(f, g) &= \overline{(g, f)}, \quad (f + g, h) = (f, h) + (g, h), \\
(\lambda f, g) &= \bar{\lambda}(g, f), \quad (f, \lambda g) = \lambda(f, g), \quad \lambda \text{ scalar}.
\end{aligned} \tag{7.7}$$

Also,

$$|(f, g)| \leq \|f\| \, \|g\|, \quad \|f + g\| \leq \|f\| + \|g\|. \tag{7.8}$$

By Lemma 7.3, f, g can be expanded into rapidly converging Fourier series of type (7.1). A fundamental property of Fourier series in relation to the L_2-norm and scalar product is

Theorem 7.4 *(Parseval's Relation) If f and g have finite L_2-norm, we have*

$$(f, g) = \sum_{\omega=-\infty}^{+\infty} \bar{\hat{f}}(\omega) \hat{g}(\omega), \quad \text{i.e.,} \quad \|f\|^2 = \sum_{\omega=-\infty}^{+\infty} |\hat{f}(\omega)|^2. \tag{7.9}$$

Proof:

$$\begin{aligned}
(f, g) &= \int_0^1 \left(\sum_{\omega=-\infty}^{+\infty} e^{-2\pi i \omega x} \bar{\hat{f}}(\omega) \right) \left(\sum_{\mu=-\infty}^{+\infty} e^{2\pi i \mu x} \hat{g}(\mu) \right) dx \\
&= \sum_{\omega=-\infty}^{+\infty} \sum_{\mu=-\infty}^{+\infty} \int_0^1 e^{-2\pi i \omega x} e^{2\pi i \mu x} dx \, \bar{\hat{f}}(\omega) \hat{g}(\mu) \\
&= \sum_{\omega=-\infty}^{+\infty} \bar{\hat{f}}(\omega) \hat{g}(\omega).
\end{aligned}$$
∎

Let $f(x)$ be a piecewise continuous 1-periodic function. Then we can calculate its Fourier coefficients

$$\hat{f}(\omega) = \int_0^1 e^{-2\pi i \omega x} f(x) \, dx.$$

For every fixed N, the partial sum

$$f_N(x) = \sum_{\omega=-N}^{N} e^{2\pi i \omega x} \hat{f}(\omega) \tag{7.10}$$

belongs to $C^\infty(-\infty, \infty)$, and by Lemma 7.3 and Theorem 7.4, expression (7.10) can be developed into a rapidly converging Fourier series, satisfying

$$\|f_N\|^2 = \sum_{|\omega| \leq N} |\hat{f}(\omega)|^2.$$

We now ask the question: Does $f_N(x)$ for $N \to \infty$ converge to $f(x)$ in the L_2-norm; that is,

$$\lim_{N \to \infty} \|f(x) - f_N(x)\|^2 = 0?$$

For our example (7.4), there is no convergence in the maximum norm, because $|\hat{g}(\omega)|$ decays only like $|\hat{g}(\omega)| \sim \text{const.}/(|\omega| + 1).^2$ However, there is convergence in the L_2-norm. We have

$$\lim_{N \to \infty} \|g(x) - g_N(x)\|^2 = \lim_{N \to \infty} \sum_{|\omega| \geq N} |\hat{g}(\omega)|^2$$

$$\leq \text{const.} \lim_{N \to \infty} \sum_{|\omega| \geq N} \frac{1}{(|\omega| + 1)^2} = 0.$$

A generalization is

Theorem 7.5 *Let $f(x)$ be 1-periodic and $\hat{f}(\omega)$ its Fourier coefficients. Assume that there are constants $K > 0$, $\delta > \frac{1}{2}$ which do not depend on ω such that*

$$|\hat{f}(\omega)| \leq \frac{K}{|\omega|^\delta + 1}.$$

Then $f(x)$ can be developed into a Fourier series and the partial sums $f_N(x)$ converge to $f(x)$ in the L_2-norm.

In abstract terms one can also express this result in the following way:

1. All 1-periodic functions $f(x) \in C^\infty(-\infty, \infty)$ belong to L_2.

2. Any other 1-periodic function $f(x)$ belongs to L_2 if there is a sequence od 1-periodic functions

$$f_j(x) \in C^\infty(-\infty, \infty)$$

such that

$$\lim_{j \to \infty} \|f_j(x) - f(x)\|^2 = 0.$$

Thus, L_2 consists of 1-periodic functions $f(x) \in C^\infty(-\infty, \infty)$ and all sequences $\{f_j(x) \in C^\infty(-\infty, \infty)\}$ of 1-periodic functions that converge in the L_2-norm.

This process is similar to the one used to define real numbers. In that case we start with rational numbers and add all sequences of rational numbers which are convergent. Of course, we calculate only with rational numbers.

We shall use the same technique and assume that the functions we deal with are C^∞-smooth. If not, we approximate them by smooth functions.

---

[2] In this example the convergence can not be uniform since the limit function is discontinuous.

7.3   Fourier interpolation

Let $f(x)$ be a smooth 1-periodic function [i.e., $f(x) = f(x+1)$ for all x]. In Section 7.1 we have shown that we can expand $f(x)$ into a fast-converging Fourier series. Here we want to show that we can interpolate the restriction of $f(x)$ on a grid by a Fourier polynomial.

Let M be a positive integer, $h = (2M + 1)^{-1}$, a mesh size that defines the grid points $x_j = jh$, $j = 0, \pm 1, \pm 2, \ldots$. We want to show that we can find interpolation coefficients $\tilde{f}(\omega)$ such that the grid function $f_j = f(x_j)$, $j = 0, \pm 1, \pm 2, \ldots$, is interpolated as

$$f_j = \sum_{\omega=-M}^{M} e^{2\pi i \omega x_j} \tilde{f}(\omega) \quad \text{for all grid points.} \tag{7.11}$$

By assumption, $f(x_j)$ is a 1-periodic function on the grid. As

$$e^{2\pi i \omega x_{j+(2M+1)}} = e^{2\pi i \omega(hj+(2M+1)h)} = e^{2\pi i \omega(x_j+1)} = e^{2\pi i \omega x_j},$$

the same is true for the exponential grid functions in (7.11) and the representation is consistent.

Now we can proceed as in Section 7.1. The interpolation coefficients $\tilde{f}(\omega)$ are determined uniquely by the grid values f_j, $j = 0, 1, 2, \ldots, 2M$. This can be seen in the same way as in Section 7.1. By (A.1) of Section A.1, the discrete version of (7.2) holds. That is, assuming $0 \leq \omega, \nu \leq 2M$, we have

$$\sum_{j=0}^{2M} h e^{2\pi i \omega x_j} e^{-2\pi i \nu x_j} = h \sum_{j=0}^{2M} \left(e^{2\pi(\omega-\nu)h} \right)^j$$

$$= \begin{cases} h(2M+1) = 1 & \text{if } \omega = \nu, \\ h\dfrac{1 - e^{2\pi i(\omega-\nu)(2M+1)h}}{1 - e^{2\pi i(\omega-\nu)h}} = 0 & \text{if } 0 < |\omega - \nu| \leq 2M. \end{cases} \tag{7.12}$$

Let μ be fixed. Then (7.11) and (7.12) give us

$$\sum_{j=0}^{2M} e^{-2\pi i \mu x_j} f(x_j) h = \sum_{\omega=-M}^{M} \left(\sum_{j=0}^{2M} e^{2\pi i(\omega-\mu)x_j} \right) \tilde{f}(\omega) h = \tilde{f}(\mu). \tag{7.13}$$

Therefore, the interpolation coefficients are given by (7.13).

Fast Fourier transform.   It is a remarkable fact that the amount of computational work required to compute the interpolation coefficients is only $\mathcal{O}(M \log M)$ elementary operations. The corresponding algorithm, which calculates the output $\tilde{f}(\omega)$, $-M \leq \omega \leq M$, from the input f_j, $0 \leq j \leq 2M$, is called the *fast Fourier transform* (FFT). The inverse transformation, which calculates the $2M + 1$ grid values f_j from the $2M + 1$ coefficients $\tilde{f}(\omega)$, can also be performed in $\mathcal{O}(M \log M)$ operations and is called the *inverse fast Fourier transform* (IFFT) (see [4, 1]).

7.3.1 Scalar product and norm for 1-periodic grid functions

Corresponding to Section 7.2, we introduce a scalar product and norm defined on the space of 1-periodic grid functions (often called l_2-scalar product and norm).

Definition 7.6 *Let* $f(x_j) = f_j$, $g(x_j) = g_j$ *be 1-periodic grid functions of type* (7.11). *The inner product and norm for this grid functions are*

$$(f, g)_h = \sum_{j=0}^{2M} \overline{f_j}\, g_j\, h,$$

$$\|f\|_h = \sqrt{(f, f)_h} = \left(\sum_{j=0}^{2M} |f_j|^2\, h \right)^{1/2}. \tag{7.14}$$

Using (7.11), (7.12) we obtain

$$\begin{aligned}
(f, g)_h &= \sum_{j=0}^{2M} \left(\sum_{\omega=-M}^{M} e^{-2\pi i \omega x_j} \overline{\tilde{f}(\omega)} \sum_{\mu=-M}^{M} e^{2\pi i \mu x_j} \tilde{g}(\mu) \right) h \\
&= \sum_{\omega=-M}^{M} \sum_{\mu=-M}^{M} \left(\sum_{j=0}^{2M} h\, e^{-(2\pi i(\omega-\mu)x_j)} \right) \overline{\tilde{f}(\omega)} \tilde{g}(\mu) \\
&= \sum_{\omega=-M}^{M} \overline{\tilde{f}(\omega)} \tilde{g}(\omega).
\end{aligned} \tag{7.15}$$

Moreover,

$$\|f\|_h^2 = \sum_{\omega=-M}^{M} |\tilde{f}(\omega)|^2. \tag{7.16}$$

Equations (7.15) and (7.16) are a discrete version of Parseval's relation.

An important property of the Fourier interpolation is that it preserves the smoothness of the original function. We defer a precise statement and proof of this result to Chapter 9, where we discuss discrete approximations to derivatives.

Exercise 7.2 *Generalize* (7.12) *to show that for all integers* ω, μ, n,

$$\sum_{j=0}^{2M} h e^{2\pi i \omega x_j} e^{-2\pi i \mu x_j} = \begin{cases} 1 & \text{if } \omega - \mu = (2M+1)n, \\ 0 & \text{otherwise.} \end{cases}$$

and therefore (7.13) *defines* $\tilde{f}(\mu)$ *for any integer* μ. *Also show that*

$$\tilde{f}(\omega) = \tilde{f}(\omega + (2M+1)n)).$$

Exercise 7.3 *Let* $g(x)$ *be the function defined in* (7.4). *Prove, using finite geometric series, that the interpolation coefficients are*

$$\tilde{g}(\omega) = -2i \frac{\sin(\pi \omega h M) \sin(\pi \omega h (M+1))}{\sin(\pi \omega h)} h. \tag{7.17}$$

Exercise 7.4 *Consider the discontinuous "sawtooth" 1-periodic function*

$$s(x) = \begin{cases} x & \text{if } -\tfrac{1}{2} < x < \tfrac{1}{2} \\ 0 & \text{if } x = \tfrac{1}{2}, \end{cases} \qquad s(x+1) = s(x),$$

in the interval $x \in [0, 1)$.

   (a) *Compute the Fourier coefficients* $\hat{s}(\omega)$.

   (b) *Using* $M = 10$, *compute and plot the truncated Fourier sum*

$$s_M^F(x) = \sum_{\omega=-M}^{M} \hat{s}(\omega) e^{2\pi i \omega x}.$$

Plot the error $e_M^F(x) = s_M^F(x) - s(x)$ *separately.*

   (c) *For* $M = 10$, *compute and plot the Fourier interpolating polynomial*

$$s_M^I(x) = \sum_{\omega=-M}^{M} \tilde{s}(\omega) e^{2\pi i \omega x}$$

and its error $e_M^I(x) = s_M^I(x) - s(x)$.

   (d) *Repeat items* (b) *and* (c) *for* $M = 100$.

CHAPTER 8

1-PERIODIC SOLUTIONS OF TIME DEPENDENT PARTIAL DIFFERENTIAL EQUATIONS WITH CONSTANT COEFFICIENTS

We start this chapter by studying three simple examples of partial differential equations and their initial value problems in the 1-periodic case. These examples are fundamental as model equations and to introduce the concepts of well posedness and stability under lower-order perturbations. At the end we generalize these concepts to more general equations and systems of equations.

8.1 Examples of equations with simple wave solutions

8.1.1 One-way wave equation

Perhaps the simplest example is the one-way wave equation

$$\frac{\partial u}{\partial t} = a \frac{\partial u}{\partial x}, \quad a \text{ real constant}, \tag{8.1}$$

with initial data:

$$u(x,0) = e^{2\pi i \omega x} \hat{f}(\omega), \quad \omega \text{ integer}. \tag{8.2}$$

Introduction to Numerical Methods for Time Dependent Differential Equations, First Edition.
By Heinz-O. Kreiss and Omar E. Ortiz. Copyright © 2014 John Wiley & Sons, Inc.

We solve the problem by separation of variables; that is, we assume that

$$u(x, t) = e^{2\pi i \omega x} \hat{u}(\omega, t). \tag{8.3}$$

By (8.2),

$$\hat{u}(\omega, 0) = \hat{f}(\omega).$$

Introducing (8.3) into (8.1) gives us

$$\frac{d\hat{u}}{dt}(\omega, t) = 2\pi i \omega a \, \hat{u}(\omega, t). \tag{8.4}$$

Therefore,

$$\hat{u}(\omega, t) = e^{2\pi i \omega a t} \hat{u}(\omega, 0),$$

and by (8.3),

$$u(x, t) = e^{2\pi i \omega (x + at)} \hat{f}(\omega). \tag{8.5}$$

We can now use the results of Section 7.2 and the principle of superposition to solve (8.1) for general initial data

$$u(x, 0) = f(x) = \sum_{\omega = -\infty}^{\infty} e^{2\pi i \omega x} \hat{f}(\omega)$$

and obtain

$$u(x, t) = \sum_{\omega = -\infty}^{\infty} e^{2\pi i \omega (x + at)} \hat{f}(\omega). \tag{8.6}$$

By Parseval's relation we obtain the *energy estimate*[1]

$$\|u(\cdot, t)\|^2 = \sum_{\omega = -\infty}^{\infty} |\hat{f}(\omega)|^2. \tag{8.7}$$

8.1.2   Heat equation

Consider now the heat equation

$$\frac{\partial u}{\partial t} = a \frac{\partial^2 u}{\partial x^2}, \quad a > 0 \text{ constant}, \tag{8.8}$$

with initial data (8.2). Again we solve the problem by the separation of variables (8.3) and now obtain

$$\frac{d\hat{u}}{dt}(\omega, t) = -4\pi^2 \omega^2 a \, \hat{u}(\omega, 0). \tag{8.9}$$

Instead of (8.5), we now obtain

$$u(x, t) = e^{2\pi i \omega x} e^{-4\pi^2 \omega^2 a t} \hat{f}(\omega). \tag{8.10}$$

---

[1] It is usual, in the context of PDE theory, to call *energy* the L_2-norm (or some other Sobolev norm) of the solution. This energy is not necessarily related to physical energy.

There is a fundamental difference between the behavior of (8.5) and (8.10). For the one-way wave equation, the amplitude of the solution does not change with time. For the heat equation, the amplitude decays rapidly with time if ω is large.

Again, by Section 7.2 and the principle of superposition, we can solve (8.1) for general initial data

$$u(x,0) = f(x) = \sum_{\omega=-\infty}^{\infty} e^{2\pi i \omega x} \hat{f}(\omega)$$

and obtain

$$u(x,t) = \sum_{\omega=-\infty}^{\infty} e^{2\pi i \omega x} e^{-4\pi^2 \omega^2 a t} \hat{f}(\omega). \tag{8.11}$$

For general initial data, the solution (8.6) has, for $t > 0$, the same smoothness properties as those of the initial data. For the heat equation, it is C^∞-smooth for $t > 0$. Therefore, it is much easier to solve (8.8) than (8.1) numerically.

Finally, corresponding to (8.7), we obtain for the heat equation an energy estimate

$$\|u(\cdot,t)\|^2 = \sum_{\omega=-\infty}^{\infty} e^{-16\pi^4 \omega^4 a^2 t^2} |\hat{f}(\omega)|^2 \le \|f(\cdot)\|^2. \tag{8.12}$$

8.1.3 Wave equation

Consider finally the wave equation,

$$\frac{\partial^2 u}{\partial t^2} = a^2 \frac{\partial^2 u}{\partial x^2}, \quad a > 0 \text{ constant.} \tag{8.13}$$

For constant initial data the solution is simply $u(x,t) = \sigma_1 + \sigma_2 t$. For nonconstant initial data we start by studying simple wave solutions. We again make the ansatz (8.3)

$$u(x,t) = e^{2\pi i \omega x} \hat{u}(\omega,t), \quad u(x,0) = \hat{u}(\omega,0), \quad \omega \ne 0,$$

and obtain a second-order ordinary differential equation

$$\frac{d^2 u}{dt^2}(\omega,t) = -a^2(4\pi^2 \omega^2)\hat{u}(\omega,t). \tag{8.14}$$

Its general solution is of the form

$$\hat{u}(\omega,t) = e^{2\pi i \omega a t} \hat{u}_1(\omega,0) + e^{-2\pi i \omega a t} \hat{u}_2(\omega,0). \tag{8.15}$$

We need two initial conditions,

$$u(x,0) = e^{2\pi i \omega x} \hat{f}_1(\omega), \quad u_t(x,0) = e^{2\pi i \omega x} \hat{f}_2(\omega). \tag{8.16}$$

Thus,

$$\hat{u}_1(\omega,0) + \hat{u}_2(\omega,0) = \hat{f}_1(\omega),$$

$$2\pi i \omega a \Big(\hat{u}_1(\omega,0) - \hat{u}_2(\omega,0) \Big) = \hat{f}_2(\omega),$$

and

$$\hat{u}_1(\omega, 0) = \frac{1}{2}\left(\hat{f}_1(\omega) + \frac{1}{2\pi i \omega a}\hat{f}_2(\omega)\right),$$
$$\hat{u}_2(\omega, 0) = \frac{1}{2}\left(\hat{f}_1(\omega) - \frac{1}{2\pi i \omega a}\hat{f}_2(\omega)\right).$$

Thus,

$$u(x, t) = e^{2\pi i \omega(x+at)}\frac{1}{2}\left(\hat{f}_1(\omega) + \frac{1}{2\pi i \omega a}\hat{f}_2(\omega)\right)$$
$$+ e^{2\pi i \omega(x-at)}\frac{1}{2}\left(\hat{f}_1(\omega) - \frac{1}{2\pi i \omega a}\hat{f}_2(\omega)\right). \quad (8.17)$$

Equation (8.17) consists of two solutions of type (8.6). We can again use the results of Section 7.2 and the principle of superposition to solve (8.13) for the general data

$$u(x, 0) = f_1(x), \quad u_t(x, 0) = f_2(x), \quad f_j(x) = \sum_{\omega=-\infty}^{\infty} e^{2\pi i \omega x}\hat{f}_j(\omega). \quad (8.18)$$

Then the general solution is

$$u(x, t) = \hat{f}_1(0) + \hat{f}_2(0)t + \sum_{\omega \neq 0}\left[e^{2\pi i \omega(x+at)}\frac{1}{2}\left(\hat{f}_1(\omega) + \frac{1}{2\pi i \omega a}\hat{f}_2(\omega)\right)\right.$$
$$\left. + e^{2\pi i \omega(x-at)}\frac{1}{2}\left(\hat{f}_1(\omega) - \frac{1}{2\pi i \omega a}\hat{f}_2(\omega)\right)\right]. \quad (8.19)$$

Again we obtain an energy estimate that, for vanishing $\hat{f}_1(0)$ and $\hat{f}_2(0)$, is

$$\|u(\cdot, t)\|^2 \leq 2\sum_{\omega \neq 0}\left(|\hat{f}_1(\omega)|^2 + \frac{1}{4\pi^2 \omega^2 a^2}|\hat{f}_2(\omega)|^2\right). \quad (8.20)$$

Exercise 8.1 *Derive estimate* (8.20).

8.2 Discussion of well posed problems for time dependent partial differential equations with constant coefficients and with 1-periodic boundary conditions

8.2.1 First-order equations

We start with

$$\frac{\partial u}{\partial t} = (a + ib)\frac{\partial u}{\partial x} + cu, \quad a, b, c \text{ real constants},$$
$$u(x, 0) = f_N(x) = \sum_{\nu=-N}^{N} e^{2\pi i \nu x}\hat{f}(\nu). \quad (8.21)$$

Because (8.21) is of the same type as the first example of Section 8.1, we could proceed as before, using the ansatz (8.3). However, we shall Fourier transform the problem directly. Since the Fourier transform of $f_N(x)$ consists of a finite number of frequencies, we know that the solution of (8.21) exists and is C^∞-smooth. We obtain

$$\int_0^1 e^{-2\pi i \omega x} \frac{\partial u}{\partial t}(x,t)\, dx = \int_0^1 e^{-2\pi i \omega x} \left((a+ib)\frac{\partial u}{\partial x}(x,t) + cu(x,t) \right) dx$$

$$\int_0^1 e^{-2\pi i \omega x} f_N(x)\, dx = \begin{cases} \hat{f}(\omega) & \text{for } |\omega| \le N \\ 0 & \text{for } |\omega| > N. \end{cases}$$

Integration by parts with respect to x, and taking the time derivative outside the integral, gives us

$$\frac{d\hat{u}}{dt}(\omega,t) = (2\pi i \omega(a+ib) + c)\,\hat{u}(\omega,t)$$

$$\hat{u}(\omega,0) = \begin{cases} \hat{f}(\omega) & \text{for } |\omega| \le N \\ 0 & \text{for } |\omega| > N; \end{cases} \tag{8.22}$$

that is, for $|\omega| \le N$,

$$\hat{u} = e^{(2\pi i \omega(a+ib)+c)t}\hat{f}(\omega). \tag{8.23}$$

The corresponding simple wave solution is given by

$$u_\omega(x,t) = e^{2\pi i \omega x} e^{(2\pi i \omega(a+ib)+c)t}\hat{f}(\omega); \tag{8.24}$$

that is,

$$|u_\omega(x,t)| = e^{(-2\pi \omega b + c)t}|\hat{f}(\omega)|. \tag{8.25}$$

Inverting the Fourier series, we obtain

$$u(x,t) = \sum_{\omega=-N}^{N} u_\omega(x,t), \tag{8.26}$$

and by Parseval's relation we obtain the energy estimate

$$\|u(x,t)\|^2 = \sum_{\omega=-N}^{N} \|u_\omega(x,t)\|^2 = \sum_{\omega=-N}^{N} e^{2(-2\pi\omega b + c)t}|\hat{f}(\omega)|^2. \tag{8.27}$$

Using the ansatz (8.3), we would have obtained the same result.

For every fixed N there is a bounded solution. However, even if ω is of moderate size, the corresponding simple wave solution can grow rapidly with time. For example, if $\omega b = -10$, $t = 10$, $c = 0$, then

$$|u_\omega(x,t)| = e^{2\pi \cdot 100}|\hat{f}(0)| \sim e^{600}|\hat{f}(0)|.$$

In numerical calculations, due to truncation errors, all frequencies are activated, that is $|\omega| \to \infty$. Therefore, we cannot calculate the solutions if $\omega b < 0$, and we call the problem *illposed*.

On the other hand, if $b = 0$, then

$$|u_\omega(x, t)| = e^{ct}|\hat{f}(\omega)|.$$

If $c > 0$, there is exponential growth but it does not depend on ω. This is quite common in applications and we have to live with it. In this case we call the problem *well posed*. The most benign situation occurs when $c \leq 0$ because there is no growth at all (see the first example in Section 8.1); we call the problem *strongly well posed*.

8.2.2 Second-order (in space) equations

Now we add a second-order term to (8.21) and consider

$$\frac{\partial u}{\partial t} = d\frac{\partial^2 u}{\partial x^2} + (a + ib)\frac{\partial u}{\partial x} + cu, \quad d \text{ real constant}, \tag{8.28}$$
$$u(x, 0) = f_N(x).$$

As before, we Fourier transform (8.28) and obtain

$$\frac{d\hat{u}}{dt}(\omega, t) = \left(-4\pi^2 d\,\omega^2 + 2\pi i\omega(a + ib) + c\right)\hat{u}(\omega, t),$$
$$\hat{u}(\omega, 0) = \begin{cases} \hat{f}(\omega) & \text{for } |\omega| \leq N \\ 0 & \text{for } |\omega| > N. \end{cases} \tag{8.29}$$

Therefore, for $|\omega| \leq N$,

$$\hat{u}(\omega, t) = e^{\left(-4\pi^2 d\,\omega^2 + 2\pi i\omega(a + ib) + c\right)t}\,\hat{f}(\omega);$$

that is,

$$|\hat{u}(\omega, t)| = e^{(-4\pi^2 d\omega^2 - 2\pi\omega b + c)t}|\hat{f}(\omega)|. \tag{8.30}$$

The behavior of the simple wave solution is determined by d.

1. If $d < 0$, for large ω,

$$e(\omega) = -4\pi^2 d\,\omega^2 - 2\pi\omega b + c \geq \text{const.} \; \omega^2 \gg 0.$$

Therefore, the exponential growth rate in (8.30) is unbounded for large $|\omega|$ and the problem is not well posed. In this case the differential equation (8.28) is called the *backward heat equation*.

2. If $d > 0$, for sufficiently large $|\omega|$ the exponent in (8.30) becomes negative:

$$e(\omega) = -4\pi^2 d\omega^2 - 2\pi\omega b + c < 0.$$

Even when $e(\omega)$ may be positive for some frequencies—for example, if $0 < d \ll 1$, we have $e(\omega) \le E$, E independent of ω. There may be exponential growth for low frequencies, but this growth cannot become arbitrarily large; that is,

$$|\hat{u}(\omega, t)| \le e^{Et}|\hat{f}(\omega)|.$$

In this case the problem is well posed.

8.2.3   General equation

We summarize our results. Consider

$$\frac{\partial u}{\partial t} = P\left(\frac{\partial}{\partial x}\right)u,$$

$$u(x, 0) = f_N(x) = \sum_{\nu=-N}^{N} e^{2\pi i \nu x} \hat{f}(\nu). \qquad (8.31)$$

Here

$$P\left(\frac{\partial}{\partial x}\right) = \sum_{j=1}^{n} a_j \frac{\partial^j}{\partial x^j}, \quad a_j \text{ complex constants.}$$

We solve the problem by Fourier transform. Since

$$\frac{\partial^j e^{2\pi i \omega x}}{\partial x^j} = (2\pi i \omega)^j e^{2\pi i \omega x},$$

we obtain

$$\hat{u}(\omega, t) = P(2\pi i \omega)\hat{u}(\omega, t),$$

$$\hat{u}(\omega, 0) = \begin{cases} \hat{f}(\omega) & \text{for } |\omega| \le N \\ 0 & \text{for } |\omega| > N. \end{cases}$$

Here $P(2\pi i \omega) = \sum_{j=1}^{n} a_j (2\pi i \omega)^j$ is called the symbol of the differential operator $P(\partial/\partial x)$. The corresponding simple wave solution is given by

$$u_\omega(x, t) = e^{2\pi i \omega x} e^{P(2\pi i \omega)t} \hat{f}(\omega).$$

Therefore,

$$|u_\omega(x, t)| = e^{\text{Re}\left(P(2\pi i \omega)\right)t}|\hat{f}(\omega)|.$$

By the superposition principle, the general solution of problem (8.31) is given by (8.26). The well posedness of the problem is characterized in terms of the symbol in the following theorem.

Theorem 8.1 *Problem (8.21) is ill posed if there is a sequence ω_j such that*

$$\lim_{j \to \infty} \text{Re } P(2\pi i \omega_j) = \infty.$$

It is well posed if there is a constant K such that for all ω,

$$\text{Re } P(2\pi i \omega) \le K.$$

It is strongly well posed if $K \le 0$.

8.2.4   Stability against lower-order terms and systems of equations

We consider problem (8.31) and assume that $a_n \neq 0$. Then we call $a_n \partial^n / \partial x^n$ the *principal part* of the differential operator $P(\partial / \partial x)$, and the other terms, *lower-order terms*. By Theorem 8.1 it is clear that for our model problems (8.21) and (8.28), the lower-order terms have no influence as to whether the problems are well posed.

To generalize this result to hyperbolic and parabolic systems of equations we will use energy estimates. Before considering systems we want to show how to get energy estimates for the one-way wave and the heat equations without using the Fourier representation. We construct energy estimates in the physical space by applying *integration by parts*, a powerful technique that can also be applied to nonperiodic problems.

One-way wave equation.   We again consider equation (8.1) with initial data $u(x, 0) = f(x)$, with f a smooth 1-periodic function. The energy is defined as

$$E(t) = \|u(\cdot, t)\|^2 = \int_0^1 |u(x, t)|^2 \, dx. \tag{8.32}$$

The energy is conserved in this problem. Integration by parts gives us

$$\begin{aligned}
\frac{d}{dt} E(t) &= \frac{d}{dt} \int_0^1 |u(x, t)|^2 \, dx \\
&= \int_0^1 \frac{\partial \bar{u}}{\partial t}(x, t) u(x, t) \, dx + \int_0^1 \bar{u}(x, t) \frac{\partial u}{\partial t}(x, t) \, dx \\
&= -a \int_0^1 \frac{\partial \bar{u}}{\partial x}(x, t) u(x, t) \, dx - a \int_0^1 \bar{u}(x, t) \frac{\partial u}{\partial x}(x, t) \, dx \\
&= -a\bar{u}(x, t) u(x, t)\big|_0^1 + a \int_0^1 \bar{u}(x, t) \frac{\partial u}{\partial x}(x, t) \, dx \\
&\quad - a \int_0^1 \bar{u}(x, t) \frac{\partial u}{\partial x}(x, t) \, dx = 0,
\end{aligned}$$

where the first term vanished because of the periodicity. Therefore, by integrating in time we get

$$\|u(\cdot, t)\|^2 = E(t) = E(0) = \|f(\cdot)\|^2, \tag{8.33}$$

which is equivalent to (8.7).

Heat equation.   Now consider equation (8.8) with initial data $u(x, 0) = f(x)$, smooth, and 1-periodic. The energy is again defined by (8.32) and decays for this problem.

$$\frac{d}{dt}E(t) = \frac{d}{dt}\int_0^1 |u(x,t)|^2 \, dx$$

$$= \int_0^1 \frac{\partial \bar{u}}{\partial t}(x,t)u(x,t)\,dx + \int_0^1 \bar{u}(x,t)\frac{\partial u}{\partial t}(x,t)\,dx$$

$$= a\int_0^1 \frac{\partial^2 \bar{u}}{\partial x^2}(x,t)u(x,t)\,dx + a\int_0^1 \bar{u}(x,t)\frac{\partial^2 u}{\partial x^2}(x,t)\,dx$$

$$= 2a\frac{\partial \bar{u}}{\partial x}(x,t)u(x,t)\Big|_0^1 - 2a\int_0^1 \left|\frac{\partial u}{\partial x}(x,t)\right|^2 dx$$

$$= -2a\int_0^1 \left|\frac{\partial u}{\partial x}(x,t)\right|^2 dx \leq 0.$$

The energy strictly decays until the solution becomes constant. In general,

$$\|u(\cdot,t)\|^2 = E(t) \leq E(0) = \|f(\cdot)\|^2. \tag{8.34}$$

Systems of equations.   We now generalize the result of the preceding section to 1-periodic hyperbolic and parabolic systems in one space dimension. We consider the hyperbolic system

$$\frac{\partial \mathbf{u}}{\partial t} = A\frac{\partial \mathbf{u}}{\partial x} + B\mathbf{u}. \tag{8.35}$$

Here $\mathbf{u} = (u_1, u_2, \ldots, u_n)^T$ is a vector-valued function with complex components and A, B are constant $n \times n$ matrices with A real, nonsingular, and symmetric.

We also consider a parabolic system of the form

$$\frac{\partial \mathbf{u}}{\partial t} = D\frac{\partial^2 \mathbf{u}}{\partial x^2} + A_1\frac{\partial \mathbf{u}}{\partial x} + B_1\mathbf{u}, \tag{8.36}$$

where D is symmetric and positive definite. To begin with, we neglect lower-order terms and consider

$$\frac{\partial \mathbf{u}}{\partial t} = A\frac{\partial \mathbf{u}}{\partial x}, \quad \text{and} \quad \frac{\partial \mathbf{u}}{\partial t} = D\frac{\partial^2 \mathbf{u}}{\partial x^2}. \tag{8.37}$$

Since A, D are symmetric, there are unitary matrices U_1 and U_2 which transform, respectively, A and D to diagonal form. Therefore, we make the change of variables $\mathbf{u} = U_1\tilde{\mathbf{u}}$, $\mathbf{u} = U_2\tilde{\mathbf{u}}$, respectively, and obtain

$$\frac{\partial \tilde{\mathbf{u}}}{\partial t} = \Lambda_1\frac{\partial \tilde{\mathbf{u}}}{\partial x}, \quad \text{and} \quad \frac{\partial \tilde{\mathbf{u}}}{\partial t} = \Lambda_2\frac{\partial^2 \tilde{\mathbf{u}}}{\partial x^2}. \tag{8.38}$$

Here Λ_j, $j = 1, 2$, are real diagonal matrices and Λ_2 is positive definite. Thus, we have reduced our problems to scalar equations, and the initial value problems are well posed.

Systems (8.35) and (8.36) can, in general, not be reduced to diagonal form. Therefore, we use integration by parts.

The L_2-norm for a vector-valued function $\mathbf{u}$ is

$$\|\mathbf{u}\|^2 = \int_0^1 |\mathbf{u}(x,t)|^2 \, dx.$$

Here $| \ |$ is the Euclidean norm introduced in Chapter 6.

For (8.35),

$$\frac{d}{dt}\|\mathbf{u}\|^2 = \frac{d}{dt}(\mathbf{u},\mathbf{u}) = \left(\frac{\partial \mathbf{u}}{\partial t}, \mathbf{u}\right) + \left(\mathbf{u}, \frac{\partial \mathbf{u}}{\partial t}\right)$$
$$= \left(A\frac{\partial \mathbf{u}}{\partial x}, \mathbf{u}\right) + \left(\mathbf{u}, A\frac{\partial \mathbf{u}}{\partial x}\right) + (B\mathbf{u},\mathbf{u}) + (\mathbf{u},B\mathbf{u}).$$

Since

$$\left(A\frac{\partial \mathbf{u}}{\partial x}, \mathbf{u}\right) = \left(\frac{\partial \mathbf{u}}{\partial x}, A\mathbf{u}\right) = -\left(\mathbf{u}, A\frac{\partial \mathbf{u}}{\partial x}\right),$$
$$|(B\mathbf{u},\mathbf{u})| \le |B| \, \|\mathbf{u}\|^2,$$

where we use the induced matrix norm (see Definition 6.1), we obtain

$$\frac{d}{dt}\|\mathbf{u}\|^2 \le 2|B| \, \|\mathbf{u}\|^2,$$

and, integrating, the energy estimate follows for system (8.35).

For (8.36),

$$\frac{d}{dt}\|\mathbf{u}\|^2 = \frac{d}{dt}(\mathbf{u},\mathbf{u}) = \left(\frac{\partial \mathbf{u}}{\partial t}, \mathbf{u}\right) + \left(\mathbf{u}, \frac{\partial \mathbf{u}}{\partial t}\right)$$
$$= \left(D\frac{\partial^2 \mathbf{u}}{\partial x^2}, \mathbf{u}\right) + \left(\mathbf{u}, D\frac{\partial^2 \mathbf{u}}{\partial x^2}\right) + \left(A_1\frac{\partial \mathbf{u}}{\partial x} + B_1\mathbf{u}, \mathbf{u}\right)$$
$$+ \left(\mathbf{u}, A_1\frac{\partial \mathbf{u}}{\partial x} + B_1\mathbf{u}\right), \tag{8.39}$$

Observing that D is positive definite, integration by parts gives us

$$\left(D\frac{\partial^2 \mathbf{u}}{\partial x^2}, \mathbf{u}\right) + \left(\mathbf{u}, D\frac{\partial^2 \mathbf{u}}{\partial x^2}\right) = -2\left(D\frac{\partial \mathbf{u}}{\partial x}, \frac{\partial \mathbf{u}}{\partial x}\right) \le -2\delta \left\|\frac{\partial \mathbf{u}}{\partial x}\right\|^2. \tag{8.40}$$

Here δ is the smallest eigenvalue of D. Also,

$$\left(A_1\frac{\partial \mathbf{u}}{\partial x} + B_1\mathbf{u}, \mathbf{u}\right) + \left(\mathbf{u}, A_1\frac{\partial \mathbf{u}}{\partial x} + B_1\mathbf{u}\right) \le 2|A_1|\|\mathbf{u}\| \left\|\frac{\partial \mathbf{u}}{\partial x}\right\| + 2|B_1|\|\mathbf{u}\|^2. \tag{8.41}$$

Using (8.40) and (8.41) in (8.39), we obtain

$$\frac{d}{dt}\|\mathbf{u}\|^2 \le \left(\frac{|A_1|}{4\delta} + 2|B_1|\right)\|\mathbf{u}\|^2. \tag{8.42}$$

Integrating, we get the energy estimate for (8.36).

Exercise 8.2 *Show* (8.42).

As with scalar equations, the lower-order terms have no influence on whether the problem is well. However, they have do have an influence on whether the problem is strongly well posed.

A very general theory for systems of differential equations in any number of space dimensions has been developed in [[7], Chapter 2]. From the point of view of well posedness, for parabolic systems, symmetric hyperbolic systems, and mixed hyperbolic-parabolic systems, the lower-order terms can be neglected.

CHAPTER 9

APPROXIMATIONS OF 1-PERIODIC SOLUTIONS OF PARTIAL DIFFERENTIAL EQUATIONS

In this chapter we introduce basic finite difference operators that approximate space derivatives and spectral derivatives that are based on Fourier theory. Then the method of lines, a very useful approach that consists of approximating the space derivatives but keeping the time continuous, is explained. Under this approach the partial differential equations become systems of ordinary differential equations, and therefore we can use the results of the first part of the book to analyze the model problems.

9.1 Approximations of space derivatives

Let M be a positive integer and $h = (2M + 1)^{-1}$ a grid size. The corresponding grid points are denoted by $x_j = jh, j = 0, \pm 1, \pm 2, \ldots$. A grid function $f_j = f(x_j)$ is simply a function defined on the grid. It is called 1-periodic if

$$f_j = f(x_j) = f(jh + (2M + 1)h) = f(x_j + 1) = f_{j+2M+1} \quad \text{for all } j \in \mathbb{Z},$$

and therefore any 1-periodic grid function is well defined by its $2M + 1$ values $f_0, f_1, \ldots, f_{2M}$.

Introduction to Numerical Methods for Time Dependent Differential Equations, First Edition. **105**
By Heinz-O. Kreiss and Omar E. Ortiz. Copyright © 2014 John Wiley & Sons, Inc.

Figure 9.1   Grid.

The most common *difference operators* approximating d/dx are D_+, D_-, D_0, which are defined by

$$D_+ f_j = \frac{1}{h}(f_{j+1} - f_j),$$

$$D_- f_j = \frac{1}{h}(f_j - f_{j-1}), \tag{9.1}$$

$$D_0 f_j = \frac{1}{2h}(f_{j+1} - f_{j-1}),$$

and which are called *forward*, *backward*, and *centered difference operators*, respectively. Clearly, if f_j is 1-periodic, then $D_+ f_j$, $D_- f_j$, and $D_0 f_j$ are also 1-periodic.

Discretization error and order of accuracy.   Let f_j denote the restriction of a smooth function $f(x)$ to the grid. The discretization error is the deviation from the exact derivative of the function at a grid point when this derivative is approximated by a difference operator. As the truncation error of a difference approximation to an ordinary differential equation, it can be computed by Taylor expansion. For example, we have

$$f_{j+1} = f_j + h\frac{df}{dx}(x_j) + \frac{h^2}{2}\frac{d^2f}{dx^2}(x_j) + \frac{h^3}{6}\frac{d^3f}{dx^3}(x_j) + \mathcal{O}(h^4),$$

A similar expansion holds for f_{j-1}. These expansions give

$$D_+ f_j - \frac{df}{dx}(x_j) = \frac{1}{h}(f_{j+1} - f_j) - \frac{df}{dx}(x_j)$$

$$= \frac{h}{2}\frac{d^2f}{dx^2}x_j) + \frac{h^2}{6}\frac{d^3f}{dx^3}(x_j) + \mathcal{O}(h^3),$$

$$D_- f_j - \frac{df}{dx}(x_j) = \frac{1}{h}(f_j - f_{j-1}) - \frac{df}{dx}(x_j)$$

$$- -\frac{h}{2}\frac{d^2f}{dx^2}(x_j) + \frac{h^2}{6}\frac{d^3f}{dx^3}(x_j) + \mathcal{O}(h^3).$$

This equations mean that D_+ and D_- are first-order accurate approximations of the first derivative of f. Because

$$D_0 = \tfrac{1}{2}(D_+ + D_-),$$

we obtain from the error expansions above,

$$D_0 f_j - \frac{df}{dx}(x_j) = \frac{h^2}{6}\frac{d^3f}{dx^3}(x_j) + \mathcal{O}(h^3); \tag{9.2}$$

that is, D_0 is a second-order accurate approximation of d/dx.

The difference operator D_0 uses the values of the grid function at three consecutive points to compute the derivative at the center point; one says that D_0 has a *span* of three points. One can, of course, write difference operators that use more than three points to construct difference operators with a higher order of accuracy.

Exercise 9.1 *Deduce a five-point span centered difference approximation to the first derivative which is accurate of order 4.*

Application to Exponentials. Because we intend to use Fourier interpolation, it is important to understand the action of the difference operators on the exponential grid functions $e^{2\pi i \omega x_j}$. Because of

$$e^{2\pi i \omega x_{j+1}} = e^{2\pi i \omega h} e^{2\pi i \omega x_j},$$

we obtain

$$
\begin{aligned}
D_+ e^{2\pi i \omega x_j} &= \frac{1}{h}(e^{2\pi i \omega h} - 1)e^{2\pi i \omega x_j} \\
&= e^{\pi i \omega h}\frac{1}{h}(e^{\pi i \omega h} - e^{-\pi i \omega h})e^{2\pi i \omega x_j} \\
&= 2i e^{\pi i \omega h}\frac{\sin(\pi \omega h)}{h}e^{2\pi i \omega x_j},
\end{aligned}
\tag{9.3}
$$

$$
\begin{aligned}
D_- e^{2\pi i \omega x_j} &= \frac{1}{h}(1 - e^{-2\pi i \omega h})e^{2\pi i \omega x_j} \\
&= e^{-\pi i \omega h}\frac{1}{h}(e^{\pi i \omega h} - e^{-\pi i \omega h})e^{2\pi i \omega x_j} \\
&= 2i e^{-\pi i \omega h}\frac{\sin(\pi \omega h)}{h}e^{2\pi i \omega x_j},
\end{aligned}
\tag{9.4}
$$

$$
\begin{aligned}
D_0 e^{2\pi i \omega x_j} &= \frac{1}{2h}(e^{2\pi i \omega h} - e^{-2\pi i \omega h})e^{2\pi i \omega x_j} \\
&= i\frac{\sin(2\pi \omega h)}{h}e^{2\pi i \omega x_j}.
\end{aligned}
\tag{9.5}
$$

We note that the factor multiplying the grid function $e^{2\pi i \omega x_j}$ on the right-hand sides is always independent of j. Therefore, the grid function $e^{2\pi i \omega x_j}$ is an *eigenfunction* of each of the operators D_+, D_-, D_0. Upon application of any of these operators, the exponential goes over into a multiple of itself.

For fixed x_j and $j \to 0$, the left-hand sides of the formula above converge to $(d/dx)e^{2\pi i \omega x}|_{x=x_j}$. This follows from the above consideration of the discretization error. Also, if we expand the sine function about zero, we see that the right-hand sides converge to $2\pi i \omega e^{2\pi i \omega x_j}$. Of course, the limits of the two sides are equal since

$$\frac{d}{dx}e^{2\pi i \omega x} = 2\pi i \omega e^{2\pi i \omega x}.
\tag{9.6}$$

Remark 9.1 *Equation* (9.6) *says that* $e^{2\pi i \omega x}$ *is an eigenvector of the operator* d/dx *with eigenvalue* $2\pi i \omega$. *If one replaces* d/dx *by* D_+ *or* D_- *or* D_0, *one obtains the* *eigenrelations (9.3)–(9.5). A remarkable fact is that only the eigenvalue gets perturbed by this replacement whereas the eigenfunction of each of the discrete operators is precisely the restriction of $e^{2\pi i\omega x}$ to the grid. We will see below why this is important for stability analysis.*

Approximation for d^2/dx^2.   A very commonly used difference approximation for the second derivative is given by D_+D_-; this is

$$D_+D_-f_j = \frac{1}{h}D_-(f_j - f_{j-1}) = \frac{1}{h^2}(f_{j+1} - 2f_j + f_{j-1}).$$

By Taylor expansion,

$$D_+D_-f_j - \frac{d^2f}{dx^2}(x_j) = \frac{h^2}{12}\frac{d^4f}{dx^4}(x_j) + \mathcal{O}(h^4), \tag{9.7}$$

which shows that D_+D_- is a second-order accurate approximation of d^2/dx^2. Also, from the expression above for D_+ and D_- applied to $e^{2\pi i\omega x_j}$, we obtain

$$D_+D_-e^{2\pi i\omega x_j} = 2ie^{-\pi i\omega h}\frac{\sin(\pi\omega h)}{h}D_+e^{2\pi i\omega x_j}$$

$$= -4\frac{\sin^2(\pi\omega h)}{h^2}e^{2\pi i\omega x_j}. \tag{9.8}$$

Exercise 9.2 *Find α such that the difference operator*

$$D^2f_j = D_+D_-(1 - \alpha h^2D_+D_-)f_j$$

is a fourth-order accurate approximation of $d^2f(x_j)/dx^2$. What is the span of this operator?

9.1.1   Smoothness of the Fourier interpolant

Here we state and prove an important property of the Fourier interpolant introduced in Section 7.3.

Let $f(x)$ be a 1-periodic function and $f^I(x)$ be its Fourier interpolant on a given grid $x_j = hj$, $j = 0, \pm1, \pm2, \ldots$, where $h = 1/(2M + 1)$ is the mesh size, that is,

$$f^I(x) = \sum_{\omega=-M}^{M} e^{2\pi i\omega x}\tilde{f}(\omega), \tag{9.9}$$

where the Fourier interpolant coefficients $\tilde{f}(\omega)$ are computed as in (7.13).

Theorem 9.2

$$\|D^pf^I\| \le \left(\frac{\pi}{2}\right)^p\|D_+^pf\|_h, \quad p = 0, 1, 2, \ldots, \quad D = \frac{d}{dx}. \tag{9.10}$$

We emphasize that the constant $(\pi/2)^p$ does not depend on the mesh size h.

Proof: We differentiate $f^I(x)$ p times and use the orthogonality of the exponentials (7.2) to get

$$\|D^p f^I\|^2 = \sum_{\omega=-M}^{M} |\tilde{f}(\omega)|^2 (2\pi\omega)^{2p}.$$

Applying D_+ p times to $f(x)$ at the grid point x_j we get, by (9.3),

$$(D_+^p f)(x_j) = \sum_{\omega=-M}^{M} 2ie^{-\pi i\omega h} \frac{\sin(\pi\omega h)}{h} e^{2\pi i\omega x_j} \tilde{f}(\omega),$$

so that, by (7.16),

$$\|D_+^p f\|_h^2 = \sum_{\omega=-M}^{M} |2\frac{\sin(\pi\omega h)}{h}|^{2p} |\tilde{f}(\omega)|^2$$

$$= \sum_{\omega=-M}^{M} |\frac{\sin(\pi\omega h)}{\pi\omega h}|^{2p} (2\pi\omega)^{2p} |\tilde{f}(\omega)|^2$$

$$\geq \sum_{\omega=-M}^{M} \left(\frac{2}{\pi}\right)^{2p} (2\pi\omega)^{2p} |\tilde{f}(\omega)|^2 = \left(\frac{2}{\pi}\right)^{2p} \|D^p f^I\|^2. \qquad \blacksquare$$

9.2   Differentiation of Periodic Functions

If we have a 1-periodic smooth function $f(x)$ we have a Fourier series representation

$$f(x) = \sum_{\omega=-\infty}^{\infty} \hat{f}(\omega)e^{2\pi i\omega x}, \tag{9.11}$$

and we know that the derivative of $f(x)$ can be obtained by multiplying every Fourier coefficient in (9.11) by $2\pi i\omega$.

Now, assume that we know the restriction of this smooth function to a grid x_j [i.e., we know the grid function $f_j = f(x_j)$]. How can we efficiently compute a good approximation to the derivative df/dx at the grid points x_j?

A good answer is: We first use FFT to calculate the coefficients $\tilde{f}(\omega)$ of the Fourier interpolant

$$f_M^F(x) = \sum_{\omega=-M}^{M} \tilde{f}(\omega)e^{2\pi i\omega x}.$$

We know that $f_j = f^I(x_j)$. Then we differentiate $f_M^F(x)$ exactly and evaluate on the grid:

$$\frac{df^I}{dx}(x_j) = 2\pi i \sum_{\omega=-M}^{M} \tilde{f}(\omega)\omega e^{2\pi i\omega x_j}.$$

The values of $df^I(x_j)/dx$ are good approximations of $df(x_j)/dx$ and can be computed efficiently by applying IFFT to the coefficients $2\pi i w \tilde{f}(\omega)$.

9.3   Method of lines

The so called *method of lines* consists of approximating the space derivatives of a partial differential equation (e.g., by difference approximations) while keeping the time continuous. This semidiscrete approach is very useful to a study of the stability of difference methods of solving the equation. In the method of lines the partial differential equation becomes a system of ordinary differential equations, as we will see. In practice, the method of lines also makes reference to a way of writing computer codes to solve partial differential equations such that the numerical method used to integrate in time can easily be changed without changing the approximation of the space derivatives. In this section we introduce the method of lines by applying it to the model equations of Chapter 8.

9.3.1   One-way wave equation

Consider the initial value problem

$$\frac{\partial u}{\partial t}(x,t) = a\frac{\partial u}{\partial x}(x,t), \quad a \text{ real constant,}$$
$$u(x,0) = f(x). \tag{9.12}$$

We assume that $f(x)$ and therefore also $u(x,t)$ are smooth functions which are 1-periodic in x. We shall only discretize the space dimension but keep the time continuous. Let M be a positive integer, $h = (2M+1)^{-1}$ a mesh size thaht defines the grid points x_j, $j = 0, \pm 1, \pm 2, \ldots$. We approximate (9.12) on the grid by

$$\frac{dv}{dt}(x_j,t) = aD_0v(x_j,t), \quad x_j = jh,$$
$$v(x_j,0) = f(x_j). \tag{9.13}$$

We are interested in solutions that are 1-periodic on the grid [i.e., $v(x_{j+2M+1},t) - v(x_j,t)$], so that (9.13) is a system of $2M+1$ coupled differential equations. The easiest way to calculate the solution is to use essentially the same procedure as that used in Chapter 8.

We start with simple wave solutions and use separation of variables, that is, given a frequency ω, we make the ansatz

$$v(x_j,t) = \tilde{v}(\omega,t)e^{2\pi i\omega x_j},$$
$$\tilde{v}(\omega,0) = \tilde{f}(\omega). \tag{9.14}$$

By Section 9.1, introducing (9.14) into (9.13) gives us

$$\frac{dv}{dt}(x_j, t) = \frac{d\tilde{v}}{dt}(\omega, t)e^{2\pi i\omega x_j}$$
$$= a\tilde{v}(\omega, t)D_0 e^{2\pi i\omega x_j} = \tilde{v}(\omega, t)q(\omega)e^{2\pi i\omega x_j}. \qquad (9.15)$$

Here $q(\omega) = ai\big(\sin(2\pi\omega h)/h\big)$ is called the *amplification factor*. Therefore,

$$\frac{d\tilde{v}}{dt}(\omega, t) = \tilde{v}(\omega, t)q(\omega), \qquad (9.16)$$

that is,

$$\tilde{v}(\omega, t) = \tilde{v}(\omega, 0)e^{q(\omega)t}. \qquad (9.17)$$

By (9.14) and (9.17),
$$v(x_j, t) = \tilde{f}(\omega)e^{2\pi i\omega x_j + q(\omega)t}, \qquad (9.18)$$

which is a second-order accurate approximation of $u(x_j, t)$ on the grid when the initial data function is $f(x_j) = \tilde{f}(\omega)e^{2\pi i\omega x_j}$. As $q(\omega)$ is purely imaginary, then $|v(x, t)| = |\tilde{f}(\omega)|$ and the approximation is stable.

It is important to notice that every mode (i.e. the solution for a fixed frequency ω), satisfies an ordinary differential equation. By (9.13)–(9.18),

$$\frac{dv}{dt}(x_j, t) = q(\omega)e^{2\pi i\omega x_j + q(\omega)t}\tilde{f}(\omega) = q(\omega)v(x_j, t). \qquad (9.19)$$

One can compare (9.18) with the exact solution (8.5). For low modes (i.e., $|\omega h| \ll 1$), $q(\omega) \simeq 2\pi i\omega a$, and the solution behaves like the exact solution. For high modes, $|\omega h| \simeq \frac{1}{2}$, and $q(\omega) \simeq 0$, the amplitude is preserved, but the *phase error* becomes large.

As our problem is linear, we can now use the results of Section 7.3 and the principle of superposition to solve (9.12) for general initial data. Consider the semi-discretized problem (9.13) with general 1-periodic initial data. The initial data function can be Fourier interpolated on the grid as in (7.11):

$$f(x_j) = \sum_{\omega=-M}^{M} \tilde{f}(\omega)e^{2\pi i\omega x_j}. \qquad (9.20)$$

Also, the solution can be Fourier interpolated:

$$v(x_j, t) = \sum_{\omega=-M}^{M} \tilde{v}(\omega, t)e^{2\pi i\omega x_j}. \qquad (9.21)$$

By orthogonality of the exponentials [see (7.12)], $\tilde{v}(\omega, 0) = \tilde{f}(\omega)$. Inserting (9.21) into (9.13) gives, again by orthogonality of the exponentials, exactly the problem

(9.16) for each individual frequency ω. By (9.21) the general solution is a superposition of solutions (9.18),

$$v(x_j, t) = \sum_{\omega=-M}^{M} \tilde{f}(\omega)e^{2\pi i \omega x_j + q(\omega)t}, \qquad (9.22)$$

which is a second-order accurate approximation of $u(x_j, t)$ on the grid. By the discrete Parseval's relation (7.16), and because $q(\omega)$ is purely imaginary,

$$\|v(\cdot, t)\|_h^2 = \sum_{\omega=-M}^{M} |\tilde{f}(\omega)e^{q(\omega)t}|^2 = \sum_{\omega=-M}^{M} |\tilde{f}(\omega)|^2 = \|f\|_h^2. \qquad (9.23)$$

By (9.23) the *energy* is preserved during time evolution by the semi-discretized equation (i.e., the approximation is stable).

We now consider the one-way wave equation (9.12) with a lower-order term,

$$\frac{\partial u}{\partial t}(x, t) = a\frac{\partial u}{\partial x}(x, t) + cu(x, t), \quad a, c \text{ real}, \qquad (9.24)$$

which we approximate on the grid by

$$\frac{dv}{dt}(x_j, t) = aD_0 v(x_j, t) + cv(x_j, t).$$

Then the amplification factor becomes

$$q(\omega) = ai\frac{\sin(2\pi\omega h)}{h} + c.$$

Thus,

$$|v(x_j, t)| = e^{ct}|f(\omega)|$$

and $v(x, t)$ behaves like the solutions in Chapter 8.

Artificial dissipation. The addition of a dissipative term to the discrete scheme (9.13) can be used to minimize the bad effect of inaccurately computed high-frequency modes. Instead of (9.13), consider the approximation

$$\frac{dv}{dt}(x_j, t) = (aD_0 + bhD_+D_-)v(x_j, t) + cv(x_j, t), \qquad (9.25)$$

which is consistent with (9.24) since the added term is proportional to a second derivative multiplied by h, and thus it vanishes formally when $h \to 0$. The amplification factor becomes

$$q(\omega) = ai\frac{\sin(2\pi\omega h)}{h} - 4b\frac{\sin^2(\pi\omega h)}{h} + c. \qquad (9.26)$$

For small $|\omega h|$ we obtain in first approximation

$$q(\omega) \sim ai2\pi\omega - 4b\pi^2\omega^2 h + c.$$

If $\omega^2 h \to 0$, then $q \to a2\pi i\omega$ and $v(x,t)$ behaves like the solution of the well-posed problem [see (8.4].

If $\omega^2 h \gg 1$ (i.e., we consider high frequencies), $\mathrm{Re}\, q \to \infty$ if $b < 0$ and $\omega^2 h \to \infty$ and the approximation is unstable. If $b > 0$, $\mathrm{Re}\, q \to -\infty$ for high frequencies and the approximation is stable.

In applications, high frequencies (i.e., if $\omega \geq 1/h$) are not calculated with any accuracy; the *artificial dissipation* term hbD_+D_-v, $b > 0$, helps to control these inaccurate high-frequency terms. Unfortunately, the approximation becomes only first-order accurate because the low frequencies are also affected. To prevent loss of accuracy, the standard artificial dissipation term is the fourth order term,

$$-bh^2 D_+^2 D_-^2 v(x_j, t), \quad b > 0.$$

Now the approximation is second-order accurate.

Exercise 9.3 *Prove that the approximation*

$$\frac{dv}{dt}(x_j, t) = (aD_0 - bh^2 D_+^2 D_-^2)v(x_j, t) + cv(x_j, t), \quad b > 0,$$

to (9.24) is second-order accurate.

The purpose of the method of lines is to have a way to sort out the methods thaht are definitely unstable. For equations with variable coefficients, one uses the *principle of frozen coefficients*. For example, if in (9.12) $a = a(x, t)$ is a smooth function, one replaces $a(x, t)$ by $a(x_0, t_0)$ and treats the constant coefficient problem. If, for any (x_0, t_0), the method is unstable, the method has to be improved. General theory tells us that if the method is stable at $(x, t) = (x_0, t_0)$, it must be stable in the neighborhood of that point. If a method is stable at all relevant values of (x, t), it can be applied to the variable coefficient problem.

9.3.2 Heat equation

Consider the 1-periodic initial value problem for the heat equation as in Chapter 8:

$$\frac{\partial u}{\partial t} = a\frac{\partial^2 u}{\partial x^2}, \quad a > 0, \tag{9.27}$$

$$u(x, 0) = f(x), \quad f(x + 1) = f(x).$$

On the same grid as for the one-way wave equation, we approximate (9.27) by

$$\frac{dv}{dt}(x_j, t) = aD_+D_-v(x_j, t), \tag{9.28}$$

$$v(x_j, 0) = f(x_j),$$

and look for 1-periodic solutions on the grid. With the simple wave ansatz (9.14), we again obtain

$$v(x_j, t) = \tilde{f}(\omega)e^{2\pi i\omega x_j + q(\omega)t}, \tag{9.29}$$

where the amplification factor is now

$$q(\omega) = -4a\frac{\sin^2(\pi\omega h)}{h^2}.$$

(9.30)

This is as in the previous case a second-order accurate approximation of $u(x_j, t)$ on the grid when the initial data function is $f(x_j) = \tilde{f}(\omega)e^{2\pi i\omega x_j}$.

All the approximate modes (9.29) with $\omega \neq 0$ are decaying exponentially with time, so that the approximation is stable. Notice that $\omega h \in (-\frac{1}{2}, \frac{1}{2})$. For $|\omega h|$ small,

$$q(\omega) \sim -4\pi^2 a\omega^2$$

and the solution is dominated by the exact solution (8.10). When $|\omega h| \sim \frac{1}{2}$, the approximation is not accurate, but this is not important, as both the exact and approximate modes decay, for any $t > 0$, very fast exponentially.

Each individual mode (9.29) satisfies an ordinary equation

$$\frac{dv}{dt}(x_j, t) = q(\omega)v(x_j, t).$$

(9.31)

In complete analogy with the one-way equation, the solution of (9.27) with general 1-periodic initial data (9.20) is obtained, by superposition of simple wave solutions as in (9.22) but with $q(\omega)$ given by (9.30). Using the discrete Parseval's relation, one can prove that the energy $E(t) = \|v(x_j, t)\|^2$ decays to $|\tilde{f}(0)|^2$; thus, the approximation is stable.

We consider now the heat equation perturbed with lower-order terms [see eq. (8.28)]. We study the second-order accurate semidiscrete approximation

$$\frac{dv}{dt}(x_j, t) = dD_+D_-v(x_j, t) + (a + ib)D_0v(x_j, t) + cv(x_j, t), \quad d > 0.$$

Fourier transforming, we find that

$$\frac{d\tilde{v}}{dt}(\omega, t) = q(\omega)\tilde{v}(\omega, t),$$

where the amplification factor is

$$q(\omega) = -4d\frac{\sin^2(\pi\omega h)}{h^2} + (a + ib)\frac{\sin(2\pi\omega h)}{h} + c.$$

Thus, when $|\omega h|$ is small, $q(\omega) \equiv -4\pi^2\omega^2 d + 2\pi\omega(a + ib) + c$, and the solution behaves like the exact solution of Chapter 8.

9.3.3   Wave equation

We consider our third model problem of Chapter 8, the 1-periodic wave equation

$$\frac{\partial^2 u}{\partial t^2} = a^2\frac{\partial^2 u}{\partial x^2}, \quad a \text{ real},$$

$$u(x, 0) = f_1(x), \quad \frac{\partial u}{\partial t}(x, 0) = f_2(x),$$

(9.32)

with $f_1(x+1) = f_1(x)$ and $f_2(x+1) = f_2(x)$. On the same grid as in the previous model problems, we approximate (9.32) by

$$\frac{d^2v}{dt^2}(x_j, t) = a^2 D_+ D_- v(x_j, t),$$

$$v(x_j, t) = f_1(x_j), \quad \frac{dv}{dt}(x_j, t) = f_2(x_j). \tag{9.33}$$

Again we look for simple wave solutions of the form

$$v(x_j, t) = \tilde{v}(\omega, t)e^{2\pi i \omega x_j},$$

$$\tilde{v}(x_j, 0) = \tilde{f}_1(\omega), \quad \frac{d\tilde{v}}{dt}(x_j, 0) = \tilde{f}_2(\omega). \tag{9.34}$$

Inserting (9.34) into (9.33) gives us

$$\frac{d^2\tilde{v}}{dt^2} = -4a^2 \frac{\sin^2(\pi\omega h)}{h^2}\tilde{v}. \tag{9.35}$$

Therefore, calling $q(\omega) = 2a[\sin(\pi\omega h)/h]$, we have

$$\tilde{v}(\omega, t) = \tilde{u}_1(\omega, 0)e^{iq(\omega)t} + \tilde{u}_2(\omega, 0)e^{-iq(\omega)t}, \tag{9.36}$$

and (9.34) gives

$$\tilde{u}_1(\omega, 0) = \frac{1}{2}\left(\tilde{f}_1(\omega) + \frac{1}{iq(\omega)}\tilde{f}_2(\omega)\right),$$

$$\tilde{u}_2(\omega, 0) = \frac{1}{2}\left(\tilde{f}_1(\omega) - \frac{1}{iq(\omega)}\tilde{f}_2(\omega)\right). \tag{9.37}$$

The simple wave solution is then

$$v(x_j, t) = \frac{1}{2}\left(\tilde{f}_1(\omega) + \frac{1}{iq(\omega)}\tilde{f}_2(\omega)\right)e^{2\pi i \omega x_j + iq(\omega)t}$$

$$+ \frac{1}{2}\left(\tilde{f}_1(\omega) - \frac{1}{iq(\omega)}\tilde{f}_2(\omega)\right)e^{2\pi i \omega x_j - iq(\omega)t}. \tag{9.38}$$

This solution is a second-order accurate approximation of the exact simple wave solution (8.17). When ωh is small, $q(\omega) \sim 2a\pi\omega$ and the (9.38) behaves like the exact solution. When $|h\omega|$ is large, $|h\omega| \sim \frac{1}{2}$, we have $q(\omega) \sim 4a\omega$ compared to the value $2\pi a\omega$ of the exact solution (i.e., the phase error of high modes is large). One can improve the situation by using a higher-order approximation.

9.4 Time Discretizations and Stability Analysis

One-way wave equation.   We shall now also discretize time. By (9.19),

$$\frac{dv}{dt}(x_j, t) = q(\omega)e^{2\pi i \omega x_j + q(\omega)t}\tilde{f}(\omega) = q(\omega)v(x_j, t). \tag{9.39}$$

Thus, we have, for every grid point x_j and every frequency ω, an ordinary differential equation which we write in the form

$$\frac{dy}{dt} = q(\omega)y(t). \tag{9.40}$$

Assume that we want to integrate in time using a time step k, that is, we discretize time as $t_n = kn$, $n = 0, 1, 2, \ldots$. We need the chosen method to be stable to integrate all modes. In other words, we need $kq(\omega)$ to belong to the stability region of the method for all ω.

For wave propagation problems, the leapfrog method, which we discussed in Chapter 5, is often a very efficient method. Let v_n denote the approximation to $y(t_n)$. Then

$$\frac{v_{n+1} - v_{n-1}}{2k} = q(\omega)v_n, \quad q(\omega) \text{ purely imaginary}. \tag{9.41}$$

As $q(\omega)$ is pure imaginary, $k > 0$ can be chosen small enough so that $kq(\omega)$ belongs to the leapfrog's stability region (see Theorem 5.3). Moreover,

$$q(\omega) = i\frac{a}{h}\sin(2\pi\omega h), \quad \text{with} \quad \omega h \in \left(-\frac{1}{2}, \frac{1}{2}\right).$$

Therefore $|q(\omega)| < |a|/h$, and the condition

$$k \le \frac{h}{|a|}, \quad \text{or equivalently } |a| \le \frac{h}{k}, \tag{9.42}$$

implies stability. For condition (9.42), is known as *CFL condition*,[1] the maximum speed of propagation of the completely discretized scheme (i.e., h/k) needs to be higher or equal to the physical speed $|a|$ of the wave. CFL conditions frequently appear as stability conditions in wave propagation problems.

If $q(\omega)$ has a real part, $kq(\omega)$ cannot belong to leapfrog's stability region. One can use the modified leapfrog method of Chapter 5. Assume that

$$q(\omega) = \eta(\omega) + i\xi(\omega), \quad \eta(\omega) \le 0, \quad \eta(\omega), \xi(\omega) \text{ real};$$

then the *modified leapfrog* method is

$$(1 - k\eta(\omega))v_{n+1} = (1 + k\eta(\omega))v_{n-1} + 2ki\xi(\omega)v_n. \tag{9.43}$$

In Exercise 5.2 we found that this method is stable if $k\eta(\omega) \le 0$ and $k\xi(\omega) < 1$.

Exercise 9.4 *The approximation* (9.25) *for the constant coefficient problem* (9.12) *can be cured to be second-order accurate by tunning the parameter b if one uses explicit Euler to integrate in time. Consider the approximation* (9.25) *for problem* (9.12) *and use explicit Euler to integrate in time. Prove that the truncation error*

$$R(x_j, t) = \frac{u(x_j, t+k) - u(x_j, t)}{k} - (aD_0 + bhD_+D_-)u(x_j, t), \tag{9.44}$$

---

[1] From Courant-Friedrich-Lewy.

where $u(x,t)$ is the exact solution of (9.12), is

$$R(x_j, t) = -a\frac{h^2}{6}\frac{d^3u}{dx^3}(x_j, t) + \left(\frac{ka^2}{2} - bh\right)\frac{d^2u}{dx^2}(x_j, t)$$
$$+ \frac{k^2}{6}\frac{d^3u}{dt^3}(x_j, t) - b\frac{h^2}{12}\frac{d^4u}{dx^4}(x_j, t) + \mathcal{O}(k^3). \quad (9.45)$$

Therefore, if we choose $b = a^2k/2h$, the method becomes second-order accurate in both space and time. The approximation (9.44) is known as the Lax-Wendroff *method.*

Heat equation. If we discretize time and use, for example, the explicit Euler method to integrate the equation, stability requires that $-2 \leq kq(\omega) \leq 0$, as $-4a/h^2 \leq q(\omega) \leq 0$. Then k is chosen so that

$$k \leq \frac{h^2}{2}\frac{1}{a}.$$

Even when explicit Euler is only first-order accurate, the choice of k proportional to h^2 gives an approximation that is second-order accurate in the parameter h.

Exercise 9.5 *Consider initial value problem (9.27) with $a = 1$ and*

$$f(x) = \sin(2\pi x) + 10\sin(10\pi x), \quad x \in [0, 1).$$

(a) *Solve the problem analytically. Plot the initial data and the solution $u(x,t)$ at $t = 0.004$ and $t = 0.2$.*
(b) *Let $h = 1/N$ and consider the difference approximation*

$$\frac{v_j^{n+1} - v_j^n}{k} = D_+D_-v_j^n, \quad j = 0, 1, 2, \ldots, 1 - h, \; n = 0, 1, 2, \ldots$$

with $k = h^2/10$, where $v_j^n = v(x_j, t_n)$ approximates the solution at the grid point $x_j = hj$ and time $t_n = kn$. Write a computer program that computes v_j^n and show plots of the solution at the same times as those for item (a) using $h = 1/10$.
(c) *Plot the error $e_j^n = v_j^n - u(x_j, t_n)$, at the same times of item (a).*
(d) *Let us now denote by w the approximation to $u(x_j, t_n)$ computed with the same scheme but on a finner grid that uses $h/2$ instead of h (and $k/4$ instead of k). Compute and plot the precision quotient*

$$Q(x_j, t_n) = \frac{v(x_j, t_n) - u(x_j, t_n)}{w(x_j, t_n) - u(x_j, t_n)}$$

at $t_n = 0.2$ as a function of x_j. Do you get the resulti expected?
(e) *Repeat items (b), (c) and (d) with $h = 1/100$.*

Wave equation.   Differentiating (9.38) twice, we get for every mode

$$\frac{d^2 v}{dt^2} = -q^2(\omega)v_\omega, \tag{9.46}$$

which has the form of an ordinary differential equation,

$$\frac{d^2 y}{dt^2} = \lambda y(t). \tag{9.47}$$

Calling v^n the approximation of $y(t_n)$, the two-step method

$$\frac{v^{n+1} - v^n + v^{n-1}}{k^2} = \lambda v^n \tag{9.48}$$

is clearly second-order accurate. We want to study the stability of (9.48). As usual, we assume that $v^n = \kappa^n$; then

$$\kappa^2 - (2 + \lambda k^2)\kappa + 1 = 0. \tag{9.49}$$

If $-4 < \lambda k^2 < 0$, this equation has two different, complex, unitary roots, thus giving a stable approximation.

Exercise 9.6 *Prove the last assertion.*

In our application

$$\lambda k^2 = -q^2(\omega)k^2 = -4a^2 k^2 \frac{\sin^2(\pi\omega h)}{h^2}.$$

Without loss of generality we assume that the zero-frequency mode (constant mode) is absent, that is,

$$\int_0^1 f_1(x)\, dx = 0. \tag{9.50}$$

Then we have $0 < \sin^2(\pi\omega h) < \frac{1}{2}$, and the stability condition is satisfied if the CFL condition (9.42) is satisfied.

We have just shown that the completely discretized centered two-step approximation

$$v_j^0 = f_1(x_j),$$

$$v_j^1 = f_1(x_j) + k f_2(x_j) + a^2 \frac{k^2}{h^2}\big(f_1(x_{j+1}) - 2f_1(x_j) + f_1(x_{j-1})\big), \tag{9.51}$$

$$v_j^{n+1} = 2v_j^n - v_j^{n-1} + \frac{k^2}{2}a^2 D_+ D_- v_j^n, \qquad n = 1, 2, \ldots,$$

is second-order accurate (in both h and k) and stable if the time step is chosen to satisfy CFL condition (9.42) and (9.50) holds.

CHAPTER 10

LINEAR INITIAL BOUNDARY VALUE PROBLEMS

The purpose of this chapter is to introduce the basic ideas of initial boundary value problems for partial differential equations and their finite difference approximations. We study the model equations of Chapter 9. Energy estimates, the simplest technique to use to study well posedness for these problems, is discussed first. Then the method of lines and the energy estimates for typical semidiscrete approximations are discussed.

10.1 Well-Posed Initial Boundary Value Problems

Fourier mode analysis is not easily generalizable to problems with boundary conditions other than periodic, whereas the construction of energy estimates presents no problem. Energy estimates are thus the classical way to study well posedness of initial boundary value problems.

Introduction to Numerical Methods for Time Dependent Differential Equations, First Edition. **119**
By Heinz-O. Kreiss and Omar E. Ortiz. Copyright © 2014 John Wiley & Sons, Inc.

10.1.1   Heat equation on a strip

Here we study the initial value problem for the heat equation on a strip:

$$\frac{\partial u}{\partial t}(x,t) = a\frac{\partial^2 u}{\partial x^2}(x,t), \quad a > 0, \quad 0 \le x \le 1, \, t \ge 0,$$

$$u(x,0) = f(x),$$

(10.1)

with either fixed Dirichlet boundary conditions, $u(0,t) = U_0$, $u(1,t) = U_1$, or homogeneous Neuman boundary conditions, $\partial u(0,t)/\partial x = \partial u(1,t)/\partial x = 0$. In the first case it is convenient to redefine the solution $u(x,t)$ by subtracting the linear solution $U_0 + x(U_1 - U_0)$ from it. Then the new function satisfies the same equation but with homogeneous boundary conditions. Therefore, it is sufficient to consider problem (10.1) with either

$$u(0,t) = u(1,t) = 0$$

(10.2)

or

$$\frac{\partial u}{\partial x}(0,t) = \frac{\partial u}{\partial x}(1,t) = 0.$$

(10.3)

Smoothness of the solution.   We give an example. We want to solve (10.1), (10.2) with initial data function

$$f(x) = \begin{cases} x & \text{if } \; 0 \le x \le \frac{1}{2}, \\ 1 - x & \text{if } \; \frac{1}{2} < x \le 1. \end{cases}$$

This initial data function is only continuous and the differential equation is not satisfied at the corners of the domain $x = 0$, $t = 0$ and $x = 1$, $t = 0$.

   If we make an odd extension of f, that is,

$$\tilde{f}(x) = \begin{cases} -1 - x & \text{if } \; -1 \le x < -\frac{1}{2}, \\ x & \text{if } \; -\frac{1}{2} \le x \le \frac{1}{2}, \\ 1 - x & \text{if } \; \frac{1}{2} < x \le 1, \end{cases}$$

we can calculate the exact solution of the original problem as if it were a periodic problem for $x \in [-1,1]$ using real Fourier series, as we studied before. Thus,

$$f(x) = \sum_{n=0}^{\infty} f_n \, \sin\big((2n+1)\pi x\big), \quad f_n = \frac{2(-1)^n}{(2n+1)^2\pi^2}.$$

and the solution is

$$u(x,t) = \sum_{n=0}^{\infty} f_n e^{-a(2n+1)^2\pi^2 t} \sin\big((2n+1)\pi x\big).$$

(10.4)

For $t > 0$ the Fourier coefficients in (10.4) decay exponentially fast with n; therefore the solution is C^∞ smooth for $0 \le x \le 1$ and $t > 0$ even when the data were only continuous. This smoothness property generalizes to other boundary conditions and other strongly parabolic equations. Further details are provided elsewhere [7].

Energy estimate. We define the energy of the solution of the initial boundary value problem (10.1) and (10.2) or (10.3) by

$$E = \int_0^1 |u(x,t)|^2 \, dx. \tag{10.5}$$

In physical applications $u(x,t)$ generally represents a real variable such as temperature. However, the theory can be developed for a complex u without adding any difficulty. As some of the identities we derive are useful in other problems, we treat the case of complex $u(x,t)$.

Differentiating the energy and using the equation, we get

$$\frac{dE}{dt} = \int_0^1 \left(\frac{\partial \bar{u}}{\partial t} u + \bar{u} \frac{\partial u}{\partial t} \right) dx$$

$$= a \int_0^1 \left(\frac{\partial^2 \bar{u}}{\partial x^2} u + \bar{u} \frac{\partial^2 u}{\partial x^2} \right) dx. \tag{10.6}$$

Integrating by parts, we get

$$\frac{dE}{dt} = a \left(\frac{\partial \bar{u}}{\partial x} u + \bar{u} \frac{\partial u}{\partial x} \right) \Big|_0^1 - 2a \int_0^1 \left| \frac{\partial u}{\partial x} \right|^2 dx. \tag{10.7}$$

Any of the boundary conditions (10.2) or (10.3) say that the first term on the right-hand side vanishes. The second term is nonpositive so that

$$\frac{dE}{dt} = -2a \int_0^1 \left| \frac{\partial u}{\partial x} \right|^2 dx \le 0. \tag{10.8}$$

We obtain

$$\|u(\cdot, t)\|^2 = E(t) \le E(0) = \|f\|^2. \tag{10.9}$$

An immediate consequence of energy estimates is the *uniqueness* of the solution. If u_1 and u_2 are both solutions of the same initial boundary value problem (10.1) with (10.2) or (10.3), then by linearity the difference $u_1(x,t) - u_2(x,t)$ is also a solution of the same problem but with vanishing initial data. Then (10.9) implies that $\|u_1(\cdot, t) - u_2(\cdot, t)\|^2 = 0$; as the solutions are C^∞ smooth, they are the same.

General parabolic problem. Consider the more general parabolic problem on the strip $0 \le x \le 1$, $t \ge 0$,

$$\frac{\partial u}{\partial t} = a \frac{\partial^2 u}{\partial x^2} + b \frac{\partial u}{\partial x} + cu, \tag{10.10}$$

$$u(x,0) = f(x),$$

where $a > 0$ and $b, c \in \mathbb{C}$, with mixed boundary conditions

$$\frac{\partial u}{\partial x}(0,t) + r_0 u(0,t) = 0,$$

$$\frac{\partial u}{\partial x}(1,t) + r_1 u(1,t) = 0, \tag{10.11}$$

where r_0 and r_1 are real and f is assumed to be compatible with (10.11). In this case an energy estimate of the form

$$\|v(\cdot, t)\| \le e^{\alpha t}\|v(\cdot, 0)\|, \quad \alpha = \text{const.} \ge 0, \tag{10.12}$$

can be shown. The energy can grow exponentially, but at a rate that is independent of the initial data. We do not give a proof of this estimate here but refer the reader to reference [6].

10.1.2   One-way wave equation on a strip

We consider here the equation

$$\frac{\partial u}{\partial t} = a\frac{\partial u}{\partial x}, \quad 0 \le x \le 1, \quad t \ge 0, \tag{10.13}$$

with initial data

$$u(x, 0) = f(x), \tag{10.14}$$

and Dirichlet boundary conditions

$$\begin{aligned} (i) \quad & u(0, t) = g_0(t) \quad \text{if } a < 0, \\ (ii) \quad & u(1, t) = g_1(t) \quad \text{if } a > 0. \end{aligned} \tag{10.15}$$

Smoothness and compatibility conditions.   The smoothness of the solution of (10.13)–(10.15) depends on the smoothness of $f(x)$, $g_0(t)$, and $g_1(t)$, and also on whether these functions match smoothly at the corners $(0, 0)$ or $(1, 0)$ of the strip (i.e., the data functions satisfy the relation required by the equation at these corners).

For example, in the case $a < 0$, the data functions match smoothly up to second order at $(0, 0)$ if

$$g_0(0) = f(0),$$

$$\frac{dg_0}{dt}(0) = \frac{\partial u}{\partial t}(0, 0) = a\frac{\partial u}{\partial x}(0, 0) = a\frac{df}{dx}(0),$$

$$\frac{d^2 g_0}{dt^2}(0) = \frac{\partial^2 u}{\partial t^2}(0, 0) = a\frac{\partial^2 u}{\partial t \partial x}(0, 0)$$

$$= a\frac{\partial^2 u}{\partial x \partial t}(0, 0) = a^2\frac{\partial^2 u}{\partial x^2}(0, 0) = a^2\frac{d^2 f}{dx^2}(0).$$

We give an example. We solve the problem (10.13)–(10.15) with $a = 1$ and

$$f(x) = 0,$$

$$g_1(t) = \tfrac{1}{2}\big(1 - \cos(2\pi t)\big).$$

Here these two data functions are C^∞ smooth. However, $g_1(t)$ and $f(x)$ satisfy the equation at the right corner up to first derivatives but not to second derivatives. As a consequence, the solution

$$u(x,t) = \begin{cases} 0 & \text{if } 0 \leq x \leq 1 - t, \\ \frac{1}{2}\left(1 - \cos\left(2\pi(x - 1 + t)\right)\right) & \text{if } x > 1 - t \end{cases}$$

is C^1 smooth and not C^2 smooth for $t \geq 0$.

For the solution to be C^∞ smooth, the data functions have to match at all orders at the corners. The easiest way to achieve this is to choose C^∞ smooth initial and boundary data functions so that $f(x)$ has support on $(\varepsilon, 1 - \varepsilon)$, $0 < \varepsilon \ll 1$, and $g_i(t)$, $i = 0, 1$ vanish if $t \leq \varepsilon$. By a limit process, one can generalize this data to nonvanishing data at the corners and still get a C^∞ smooth solution. Details on this procedure are available in reference [7].

Energy estimates.   Let

$$E(t) = \int_0^1 |u(x,t)|^2 \, dx = \|u(\cdot, t)\|^2. \tag{10.16}$$

We have

Lemma 10.1 *If $u(x,t)$ solves problem (10.13)–(10.15), then*

$$E(t) \leq E(0) + |a| \int_0^t |g_i(s)|^2 \, ds,$$

with $i = 0$ if $a < 0$ and $i = 1$ if $a > 0$.

Proof: Integration by parts gives us

$$\begin{aligned}
\frac{d}{dt} E(t) &= \int_0^1 \frac{\partial \bar{u}}{\partial t}(x,t) u(x,t) \, dx + \int_0^1 \bar{u}(x,t) \frac{\partial u}{\partial t}(x,t) \, dx \\
&= a \int_0^1 \frac{\partial \bar{u}}{\partial x}(x,t) u(x,t) \, dx + a \int_0^1 \bar{u}(x,t) \frac{\partial u}{\partial x}(x,t) \, dx \\
&= a \bar{u}(x,t) u(x,t)|_0^1 - a \int_0^1 \bar{u}(x,t) \frac{\partial u}{\partial x}(x,t) \, dx + a \int_0^1 \bar{u}(x,t) \frac{\partial u}{\partial x}(x,t) \, dx \\
&= a |u(1,t)|^2 - a |u(0,t)|^2.
\end{aligned}$$

Then, in either case, $a < 0$ or $a > 0$, we neglect the negative term and replace u by the boundary data function in the positive term, obtaining an inequality. Integrating the inequality between 0 and t, the lemma follows.   ∎

Now, assume that the solution of (10.13)–(10.15) exists and is C^∞ smooth. One can differentiate the equation, initial data, and boundary condition as many times as one wants. For example, for $a > 0$,

$$\frac{\partial}{\partial t}\left(\frac{\partial u}{\partial t}\right) = a\frac{\partial^2 u}{\partial t \partial x} = a\frac{\partial}{\partial x}\left(\frac{\partial u}{\partial t}\right), \quad \frac{\partial u}{\partial t}(x,0) = a\frac{df}{dx}(x), \quad \frac{\partial u}{\partial t}(1,t) = \frac{dg_1}{dt}(t),$$

$$\frac{\partial}{\partial t}\left(\frac{\partial u}{\partial x}\right) = \frac{\partial^2 u}{\partial x \partial t} = a\frac{\partial}{\partial x}\left(\frac{\partial u}{\partial x}\right), \quad \frac{\partial u}{\partial x}(x,0) = \frac{df}{dx}(x), \quad \frac{\partial u}{\partial x}(1,t) = \frac{1}{a}\frac{dg_1}{dt}(t),$$

$$\frac{\partial}{\partial t}\left(\frac{\partial^2 u}{\partial t^2}\right) = a\frac{\partial}{\partial x}\left(\frac{\partial^2 u}{\partial t^2}\right), \quad \cdots$$

Thus, all the derivatives satisfy the same equation as u with corresponding initial and boundary conditions. Therefore, one gets estimates for all the derivatives of u which are analogous to those for u itself. These are called *a priori estimates*. For example for $\partial u/\partial t$, with $a > 0$, we have

$$\left\|\frac{\partial u}{\partial t}(\cdot,t)\right\|^2 \leq a^2\left\|\frac{df}{dx}\right\|^2 + |a|\int_0^t\left|\frac{dg_1}{dt}(s)\right|^2 ds.$$

In Section 10.2.4 we derive an energy estimate for a semidiscrete approximation of the problem (10.13)–(10.15). The existence and uniqueness of solution for the semidiscrete problem are not an issue. One can get estimates for all finite differences of the semidiscrete solution in analogy with the a priori estimates for the differential equation. These estimates can be obtained so that they are independent of the mesh size. Then one can prove that, in the limit when the mesh size goes to zero, the semidiscrete solution converges to a smooth function that is a solution of the differential equation. This is a way of showing the existence of solutions for the differential problem. Details on parabolic and hyperbolic equations are available in reference [7].

Exercise 10.1 *Prove that the solution of* (10.13)–(10.15) *is unique.*

10.1.3   Wave equation on a strip

In this section we study the wave equation by reducing it to a pair of one-way wave equations. Consider the initial value problem for the wave equation for $\varphi(x,t)$:

$$\frac{\partial^2 \varphi}{\partial t^2} = c^2\frac{\partial^2 \varphi}{\partial x^2},$$

$$\varphi(x, t = 0) = f(x), \quad \frac{\partial \varphi}{\partial t}(x, t = 0) = h(x). \tag{10.17}$$

Here $c > 0$ is the speed of propagation of the waves. If we consider the initial boundary value problem for $0 \leq x \leq 1$, we need to prescribe boundary conditions. To get the correct boundary conditions, we reduce the equation to a first-order system. Defining

$$\mathbf{u}(x,t) = \begin{pmatrix} u^1 \\ u^2 \end{pmatrix} = \begin{pmatrix} \dfrac{\partial \varphi}{\partial x} \\[2mm] \dfrac{\partial \varphi}{\partial t} \end{pmatrix}, \tag{10.18}$$

the wave equation reduces to the system

$$\frac{\partial \mathbf{u}}{\partial t} = A \frac{\partial \mathbf{u}}{\partial x}, \tag{10.19}$$

with

$$A = \begin{pmatrix} 0 & 1 \\ c^2 & 0 \end{pmatrix}. \tag{10.20}$$

The matrix A is diagonable. Calculating the transformation, we get

$$S = \frac{1}{\sqrt{2}} \begin{pmatrix} c^{-1} & -c^{-1} \\ 1 & 1 \end{pmatrix}, \quad \text{and} \quad S^{-1} = \frac{1}{\sqrt{2}} \begin{pmatrix} c & 1 \\ -c & 1 \end{pmatrix}, \tag{10.21}$$

Then, defining the variable

$$\mathbf{w} = \begin{pmatrix} w^1 \\ w^2 \end{pmatrix} = S^{-1} \mathbf{u}, \tag{10.22}$$

we obtain a diagonal system for $\mathbf{w}$:

$$\frac{\partial \mathbf{w}}{\partial t} = \Lambda \frac{\partial \mathbf{w}}{\partial x} \quad \text{with} \quad \Lambda = \begin{pmatrix} c & 0 \\ 0 & -c \end{pmatrix}, \tag{10.23}$$

with initial conditions

$$w^1(x,0) = \frac{1}{\sqrt{2}} \left(c \frac{df}{dx}(x) + h(x) \right), \quad w^2(x,0) = \frac{1}{\sqrt{2}} \left(-c \frac{df}{dx}(x) + h(x) \right). \tag{10.24}$$

The equations for w^1 and w^2 are decoupled one-way wave equations.

We know that we need to prescribe boundary conditions for w^1 at $x = 1$ and for w^2 at $x = 0$. We also want to allow coupling of the left-traveling wave with the right-traveling wave. Then the boundary conditions we consider for system (10.23) are

$$\begin{aligned} w^1(1,t) &= R^2 w^2(1,t) + \tilde{g}_1(t), \\ w^2(0,t) &= R^1 w^1(0,t) + \tilde{g}_0(t), \end{aligned} \tag{10.25}$$

where R^1, R^2 are constants, and $\tilde{g}_0(t)$, $\tilde{g}_1(t)$ are given functions, assumed to be compatible with the initial data.

We know that the problem is well posed if $R^1 = R^2 = 0$. To see whether there are restrictions on the values of R^1, R^2, we study the energy inequality. Let

$$E(t) = \int_0^1 \left(|w^1(x,t)|^2 + |w^2(x,t)|^2 \right) dx.$$

Following the proof of Lemma 10.1 and using the boundary conditions (10.25), we have

$$
\begin{aligned}
\frac{dE}{dt} &= c\big(|w^1(1,t)|^2 - |w^1(0,t)|^2\big) - c\big(|w^2(1,t)|^2 - |w^2(0,t)|^2\big) \\
&= c|R^2 w^2(1,t) + \tilde{g}_1(t)|^2 - c|w^1(0,t)|^2 - c|w^2(1,t)|^2 \\
&\quad + c|R^1 w^1(0,t) + \tilde{g}_0(t)|^2 \\
&\leq -c(1 - 2|R^2|^2))|w^2(1,t)|^2 - c(1 - 2|R^1|^2)|w^1(0,t)|^2 \\
&\quad + 2c(|\tilde{g}_0(t)|^2 + |\tilde{g}_1(t)|^2).
\end{aligned}
\tag{10.26}
$$

Therefore, if $|R^1|, |R^2| \leq \sqrt{2}/2$, we have the energy estimate

$$
E(t) \leq E(0) + 2c \int_0^t \big(|\tilde{g}_0(s)|^2 + |\tilde{g}_1(s)|^2\big)\, ds.
\tag{10.27}
$$

The boundary conditions (10.25) can, of course, be written in terms of the original variables using S^{-1}. We get

$$
\begin{aligned}
\left(c\frac{\partial\varphi}{\partial x}(1,t) + \frac{\partial\varphi}{\partial t}(1,t)\right) &= R^2\left(-c\frac{\partial\varphi}{\partial x}(1,t) + \frac{\partial\varphi}{\partial t}(1,t)\right) + g_1(t), \\
\left(-c\frac{\partial\varphi}{\partial x}(0,t) + \frac{\partial\varphi}{\partial x}(0,t)\right) &= R^1\left(c\frac{\partial\varphi}{\partial x}(0,t) + \frac{\partial\varphi}{\partial t}(0,t)\right) + g_0(t),
\end{aligned}
\tag{10.28}
$$

where $g_i(t) = \sqrt{2}\tilde{g}_i(t)$, $i = 0, 1$. These conditions can be interpreted as the *incoming wave* prescribed in terms of the *outgoing wave* and a given function at each boundary.

10.2   Method of lines

In this section we study semidiscrete approximations for the model problems of Section 10.1. The differential initial boundary value problems will be reduced to systems of ordinary differential equations which can be written in matrix form. Thus, the methods we studied in the first part of the book can be used to integrate in time.

10.2.1   Heat equation

Dirichlet boundary conditions.   Consider problem (10.1) with homogeneous Dirichlet boundary conditions (10.2). Let us call $v_j(t)$ the grid function that approximates $u(x_j, t)$ on the grid $x_j = hj$, $j = 0, 1, 2, \ldots, N-1, N$. We approximate the differential problem by

$$
\begin{aligned}
v_0(t) &= 0, \quad v_N(t) = 0, \\
\frac{dv_j}{dt} &= aD_+D_-v_j(t), \quad j = 1, 2, \ldots, N-1, \\
v_j(t = 0) &= f(x_j).
\end{aligned}
\tag{10.29}
$$

As we studied in Chapter 9, D_+D_- is a second-order accurate approximation of the second derivative, and as the boundary conditions are exact, (10.29) is a second-order accurate approximation of the original problem.

The semidiscrete problem (10.29) is an initial value problem for a system of ODEs. It can be written in matrix form as

$$\frac{d\mathbf{v}}{dt} = \frac{a}{h^2} B_D \mathbf{v}(t), \quad a > 0, \quad \mathbf{v}(0) \text{ given}, \tag{10.30}$$

where

$$\mathbf{v}(t) = \begin{pmatrix} v_1(t) \\ v_2(t) \\ \vdots \\ v_{N-1}(t) \end{pmatrix}, \quad \mathbf{v}(0) = \begin{pmatrix} f(x_1) \\ f(x_2) \\ \vdots \\ f(x_{N-1}) \end{pmatrix},$$

and

$$B_D = \begin{pmatrix} -2 & 1 & 0 & \cdots & \cdots & 0 & 0 \\ 1 & -2 & 1 & 0 & & \cdots & 0 \\ 0 & \ddots & \ddots & \ddots & & & \vdots \\ \vdots & & \ddots & \ddots & \ddots & & \vdots \\ \vdots & & & \ddots & \ddots & \ddots & 0 \\ 0 & \cdots & & 0 & 1 & -2 & 1 \\ 0 & 0 & \cdots & \cdots & 0 & 1 & -2 \end{pmatrix}.$$

Notice that the boundary conditions $v_0(t) = v_N(t) = 0$ have been used for writing this matrix representation and then these two components are absent in $\mathbf{v}(t)$.

As B_D is real and symmetric, the system of equations (10.30) is diagonable and we can apply the theory of Chapter 6. Moreover, we will show that the eigenvalues of B_D are all negative and then system (10.30) is strongly stable as a system of ODEs.

Lemma 10.2 *The matrix B_D is diagonable; its eigenvalues λ are real and lie in the interval $-4 \leq \lambda < 0$.*

Proof: The matrix B_D is real and symmetric and is therefore diagonable with real eigenvalues. We prove that all its eigenvalues are negative by showing that B_D is negative definite. Let $\mathbf{v}$ be an arbitrary vector,

$$\mathbf{v} = \begin{pmatrix} v_1 \\ v_2 \\ \vdots \\ v_{N-1} \end{pmatrix}.$$

Then, if $\mathbf{v}^t$ denotes the transpose of $\mathbf{v}$,

$$\mathbf{v}^t B_D \mathbf{v} = v_1(-2v_1 + v_2) + \sum_{i=2}^{N-2} v_i(v_{i-1} - 2v_i + v_{i+1})$$
$$+ v_{N-1}(v_{N-2} - 2v_{N-1})$$
$$= -v_1^2 - \sum_{i=2}^{N-1}(v_i - v_{i-1})^2 - v_{N-1}^2.$$

The last line is strictly negative unless $v_1 = 0$ and $v_i = v_{i-1}$, $i = 2, 3, \ldots, N - 1$ (i.e., unless $\mathbf{v}$ is the zero vector). Therefore, B_D is negative definite, and all its eigenvalues are strictly negative.

We now prove that all eigenvalues of B_D are bounded in absolute value. Let λ be any eigenvalue of B_D and $\mathbf{v}$ one of its corresponding eigenvectors, which we assume to be normalized. Then, recalling Definition 6.1, we have

$$|\lambda|^2 = |B_D \mathbf{v}|^2$$
$$= (-2v_1 + v_2)^2 + \sum_{i=2}^{N-2}(v_{i-1} - 2v_i + v_{i+1})^2 + (v_{N-2} - 2v_{N-1})^2$$
$$\leq 8v_1^2 + 2v_2^2 + \sum_{i=2}^{N-2}\left(2(v_v - v_{i-1})^2 + 2(v_{i+1} - v_i)^2\right) + 2v_{N-2}^2 + 8v_{N-1}^2$$
$$\leq 12v_1^2 + 14v_2^2 + \sum_{i=3}^{N-3} 16v_i^2 + 14v_{N-2}^2 + 12v_{N-1}^2$$
$$\leq 16|\mathbf{v}|^2 = 16.$$

Therefore, all eigenvalues of B_D lie in the interval $-4 \leq \lambda < 0$.   ∎

Lemma 10.2 says that (10.29) [or its matrix version, (10.30)] is strongly stable as a system of ordinary differential equations. We can use the explicit methods that we studied in the first part of the book to integrate this system in time. The restriction on the time step k follows from Lemma 10.2. For example, if we use explicit Euler, we need to choose the time step small enough that $-2 \leq k\lambda_j \leq 0$ for all eigenvalues λ_j of aB_D/h^2. By Lemma 10.2, this means that

$$-4\frac{a}{h^2}k \leq \lambda_j k < 0,$$

and then we need to choose k so that

$$k \leq \frac{h^2}{2a}. \tag{10.31}$$

This is the same stability condition (CFL condition) that we found in the periodic case. Other, higher-precision methods, such as Runge-Kutta methods, are of course

preferred. If we use implicit Euler method to integrate (10.30), we obtain an uncon-
ditionally stable scheme.

Exercise 10.2 *Write a computer code to compute the solution of* (10.29) *with* $a = 1$
and initial data function

$$f(x) = \begin{cases} 0 & \text{if } x = 0, \frac{1}{2}, 1 \\ 1 & \text{if } 0 < x < \frac{1}{2} \\ -1 & \text{if } \frac{1}{2} < x < 1. \end{cases}$$

(a) *Compute the solution for* $h = 1/100$ *and time step* $k = h^2/10$ *for* $t \in [0, 0.2]$.
Plot the solution at times $t = 0.002$, $t = 0.1$, *and* $t = 0.2$.

(b) *Compute two auxiliary solutions using* $h_2 = h/2$ *and* $h_3 = h/4$, *keeping the*
ratio $k_l/h_l^2 = 1/10$, *and plot the precision quotient*

$$Q(x_j, t) - \frac{v^h(x_j, t) - v^{h_2}(x_j, t)}{v^{h_2}(x_j, t) - v^{h_3}(x_j, t)}$$

at the same times as those of item (a).

Homogeneous Neuman boundary conditions.   Consider now problem (10.1) with
homogeneous Neuman boundary conditions (10.3). To be able to write a second-
order accurate semidiscrete approximation, we introduce a new grid half a step dis-
placed from the boundaries. The grid is (see Figure 10.1)

$$x_j = -\frac{h}{2} + hj, \quad j = 0, 1, \dots, N, \quad h = \frac{1}{N-1}. \tag{10.32}$$

On this grid we study the semidiscrete approximation

$$D_+ v_0(t) = 0, \quad D_- v_N(t) = 0,$$

$$\frac{dv_j}{dt} = aD_+ D_- v_j(t), \quad a > 0, \quad j = 1, 2, \dots, N-1 \tag{10.33}$$

$$v_j(t = 0) = f(x_j)$$

The one-sided difference approximations D_+ and D_- become second-order accurate
at the physical boundaries $x = 0$ and $x = 1$, respectively. The values of v_0 and v_N

Figure 10.1   Half-displaced grid with respect to the boundaries. The points x_0 and x_N are
ghost points.

(values of the solution at the *ghost points*) can be expressed in terms of v_1 and v_{N-1}, respectively, since the boundary conditions are equivalent to

$$v_0(t) = v_1(t) \quad \text{and} \quad v_N(t) = v_{N-1}(t). \tag{10.34}$$

The system of ODE (10.33) can be written as in (10.30), the only change being that the matrix B_D is now replaced by

$$
B_N = \begin{pmatrix}
-1 & 1 & 0 & \cdots & \cdots & 0 & 0 \\
1 & -2 & 1 & 0 & & \cdots & 0 \\
0 & \ddots & \ddots & \ddots & & & \vdots \\
\vdots & & \ddots & \ddots & \ddots & & \vdots \\
\vdots & & & \ddots & \ddots & \ddots & 0 \\
0 & \cdots & & 0 & 1 & -2 & 1 \\
0 & 0 & \cdots & \cdots & 0 & 1 & -1
\end{pmatrix}.
$$

This matrix has a zero eigenvalue that corresponds to the constant solution (the correct stationary solution of this problem). With minor modifications in the proof of Lemma 10.2, it can now be proved:

Lemma 10.3 *The matrix B_N is diagonable; its eigenvalues λ are real and lie in the interval $-4 \le \lambda \le 0$.*

Exercise 10.3 *Prove Lemma 10.3.*

To integrate in time, exactly the same comments as those in the Dirichlet case hold here.

10.2.2   Finite-differences algebra

As in many semidiscrete problems, it becomes difficult to analyze whether the system of ODEs is diagonable, we use energy estimates to analyze the stability of semidiscrete problems. The algebra can be presented without any extra effort for complex grid functions.

Some basic identities.   Let u_j, v_j be complex grid functions defined on any uniform grid $x_j = x_0 + hj$, $h > 0$, $j \in \mathbb{Z}$. The space of such grid functions constitutes a vector space. We introduce notation for the Euclidean inner product and norm in finite-dimensional subspaces of grid functions:

$$(u, v)_{l,m} = \sum_{j=l}^{m} h\, \bar{u}_j v_j \tag{10.35}$$

and

$$\|u\|_{l,m} = \sqrt{(u, u)_{l,m}}. \tag{10.36}$$

The following identities are a direct consequence of the definitions:

1.

$$D_+v_j = D_-v_{j+1}. \tag{10.37}$$

2. Discrete Leibnitz rules:

$$
\begin{aligned}
D_+(u_jv_j) &= (D_+u_j)v_j + u_{j+1}D_-v_{j+1}, \\
D_+(u_jv_j) &= u_jD_+v_j + (D_-u_{j+1})v_{j+1}, \\
D_-(u_{j+1}v_{j+1}) &= (D_-u_{j+1})v_j + u_{j+1}D_+v_j, \\
D_-(u_{j+1}v_{j+1}) &= (D_+u_j)v_{j+1} + u_jD_-v_{j+1}.
\end{aligned}
\tag{10.38}
$$

3. Discrete primitive:

$$
\begin{aligned}
\sum_{j=l}^{m-1} h\, D_+u_j &= u_m - u_l, \\
\sum_{l+1}^{m} h\, D_-u_j &= u_m - u_l.
\end{aligned}
\tag{10.39}
$$

Using these properties we can prove the discrete analog of integration by parts, generally called *summation by parts*.

Lemma 10.4 *Let u_j, v_j be complex grid functions defined on the grid $x_j = hj$, $j \in \mathbb{Z}$; then*

$$
\begin{aligned}
(u, D_+v)_{l,m} &= \bar{u}_{m+1}v_{m+1} - \bar{u}_l v_l - (D_-u, v)_{l+1,m+1}, \\
(D_+u, v)_{l,m} &= \bar{u}_{m+1}v_{m+1} - \bar{u}_l v_l - (u, D_-v)_{l+1,m+1}.
\end{aligned}
\tag{10.40}
$$

Proof: By the second identity in (10.38) and the first identity in (10.39), we have

$$
\begin{aligned}
(u, D_+v)_{l,m} &= \sum_{j=l}^{m} h\, \bar{u}_j D_+v_j \\
&= \sum_{j=l}^{m} h\left(D_+(\bar{u}_jv_j) - (D_-\bar{u}_{j+1})v_{j+1} \right) \\
&= \bar{u}_{m+1}v_{m+1} - \bar{u}_l v_l - (D_-u, v)_{l+1,m+1}.
\end{aligned}
$$

The proof of the second identity is completely analogous.   ∎

10.2.3   General parabolic problem

We study here a second-order-accurate semidiscrete approximation of the initial boundary value problem (10.10), (10.11). The semidiscrete system of equations approximating (10.10) is written on the grid (10.32):

$$D_+v_0 + r_0\tfrac{1}{2}(v_0 + v_1) = 0, \quad D_-v_N + r_1\tfrac{1}{2}(v_N + v_{N-1}) = 0,$$

$$\frac{dv_j}{dt} = aD_+D_-v_j + bD_0v_j + cu, \quad j = 1, 2, \ldots, N-1, \tag{10.41}$$

$$v_j(0) = f_j.$$

The boundary conditions in (10.41) are second-order-accurate approximations of the differential boundary conditions.

As in the homogeneous Neuman case, the values of v_0 and v_N (at the *ghost points*) can be expressed in terms of v_1 and v_{N-1}, respectively, by means of the boundary conditions. We have

$$v_0 = \frac{1 + hr_0/2}{1 - hr_0/2}v_1 \quad \text{and} \quad v_N = \frac{1 - hr_1/2}{1 + hr_1/2}v_{N-1}.$$

Choosing h small so that $|hr_0| \le 1$ and $|hr_1| \le 1$, we have

$$|v_0| \le 3|v_1| \quad \text{and} \quad |v_N| \le 3|v_{N-1}|. \tag{10.42}$$

We want to construct the energy estimate for the semidiscrete energy:

$$E(t) = \|v\|_{1,N-1}^2. \tag{10.43}$$

We will need a bound for the grid values v_j in terms of the norm of the solution. Clearly, $|v_j| \le h^{-1}\|v\|_{1,N-1}$, but this bound is not useful because it explodes as $h \to 0$. The useful bound involves both the norm of the grid function and the norm of its "grid derivative." We have

Lemma 10.5 *For any grid function and every $\varepsilon > 0$, we have*

$$\max_{0 \le j \le N} |f_j| \le \varepsilon\|D_-f\|_{1,N}^2 + C(\varepsilon)\|f\|_{0,N}^2, \tag{10.44}$$

where $C(\varepsilon) = 1 + \varepsilon^{-1}$.

Proof: If f_j is constant, the lemma is trivial. Assume that f_j is not constant and let l and m be indices such that

$$|f_l| = \min_{0 \le j \le N} |f_j|, \quad |f_m| = \max_{0 \le j \le N} |f_j|,$$

As f_j is not constant we can assume, for simplicity, that $l < m$. By Lemma 10.4,

$$(f, D_+f)_{l,m-1} = -(D_-f, f)_{l+1,m} + |f_j|^2|_l^m,$$

that is,

$$\max_{0 \le j \le N} |f_j|^2 \le \min_{0 \le j \le N} |f_j|^2 + \|f\|_{l,m}(\|D_+f\|_{l,m-1} + \|D_-f\|_{l+1,m})$$

$$\le \|f\|_{0,N}^2 + 2\|f\|_{0,N}\|D_-f\|_{1,N}$$

$$\le \varepsilon\|D_-f\|_{1,N}^2 + (1 + \varepsilon^{-1})\|f\|_{0,N}^2. \qquad \blacksquare$$

We also need a summation-by-parts formula for the centered derivative D_0.

Lemma 10.6

$$(u, D_0 v)_{r,s} = -(D_0 u, v)_{r,s} + \tfrac{1}{2}(\bar{u}_j v_{j+1} + \bar{u}_{j+1} v_j)|_{r-1}^s. \tag{10.45}$$

Exercise 10.4 *Prove Lemma 10.6.*

We can now estimate the energy (10.43). We have

$$
\begin{aligned}
\frac{dE}{dt} &= \left(\frac{dv}{dt}, v\right)_{1,N-1} + \left(v, \frac{dv}{dt}\right)_{1,N-1} \\
&= a\big((D_+ D_- v, v)_{1,N-1} + (v, D_+ D_- v)_{1,N-1}\big) \\
&\quad + b\big((D_0 v, v)_{1,N-1} + (v, D_0 v)_{1,N-1}\big) + 2c\|v\|_{1,N-1}^2.
\end{aligned} \tag{10.46}
$$

Applying Lemmas 10.4 and 10.6 and the boundary conditions, we have

$$
\begin{aligned}
\frac{dE}{dt} &= -2a\|v\|_{1,N-1}^2 - ar_1\left(|v_1|^2 + \frac{\bar{v}_{N-1}v_N + \bar{v}_N v_{N-1}}{2}\right) \\
&\quad + ar_0\left(|v_0|^2 + \frac{\bar{v}_0 v_1 + \bar{v}_1 v_0}{2}\right) \\
&\quad + \frac{b}{2}\left(\bar{v}_{N-1}v_N + \bar{v}_N v_{N-1} - \bar{v}_0 v_1 - \bar{v}_1 v_0\right) + 2c\|v\|_{1,N-1}^2.
\end{aligned} \tag{10.47}
$$

Using (10.42) first and then (10.44)

$$
\begin{aligned}
\frac{dE}{dt} &\leq -2a\|v\|_{1,N-1}^2 + (4a|r_1| + 3|b|)|v_{N-1}|^2 + (4a|r_0| + 3|b|)|v_1|^2 \\
&\quad + 2|c|\|v\|_{1,N-1}^2 \\
&\leq -2a\|v\|_{1,N-1}^2 + \big(4a(|r_0| + |r_1|) + 3|b|\big)\varepsilon\|D_- v\|_{2,N}^2 \\
&\quad + \big(4a(|r_0| + |r_1|) + 3|b|\big)C(\varepsilon)\|v\|_{1,N-1}^2 + 2|c|\|v\|_{1,N-1}^2.
\end{aligned} \tag{10.48}
$$

Choosing ε so that $-2a + \big(4a(|r_0| + |r_1|) + 3|b|\big)\varepsilon \leq 0$, we finally have

$$\frac{dE}{dt} \leq \alpha\|v\|_{1,N-1}^2 = \alpha E, \tag{10.49}$$

where $\alpha = \big(4a(|r_0| + |r_1|) + 3|b|\big)C(\varepsilon) + 2|c|$ is a constant independent of h, the initial data. The energy estimate follows:

$$E(t) \leq e^{\alpha t} E(0). \tag{10.50}$$

The semidiscrete problem behaves in the same way as the differential case (10.12); the energy can grow with time, but at a rate that is independent of the initial data. One can use explicit Euler, Runge-Kutta, implicit Euler, or the trapezoidal rule (see Exercise 4.2) to integrate in time.

10.2.4 One-way wave equation

We study here a second-order-accurate semidiscrete approximation of the problem (10.13)–(10.15). To this end we again use the grid (10.32) (see Figure 10.1), with two ghost points outside the physical boundaries. The cases $a > 0$ and $a < 0$ are completely analogous. Here we present the case $a > 0$.

We call $v_j(t)$ the approximation of $u(x_j, t)$. To have a second order accurate semidiscrete approximation, we use D_0 in the interior of the domain. Then, we need to complete the system with equations for the values at the ghost points x_0 and x_N. The value of v_N is determined at all times by the boundary condition, which is second order accurate, at $x = 1$. For v_0 we extend the initial data and evolution equation using the first order accurate operator D_+ so that we get a stable scheme.[1]

$$v_j(0) = f(x_j), \quad j = 1, 2, \ldots, N-1, \quad v_0(0) = v_1(0),$$

$$\frac{dv_j}{dt} = aD_0 v_j, \quad j = 1, 2, \ldots, N-1, \tag{10.51}$$

$$\frac{dv_0}{dt} = aD_+ v_0, \quad \tfrac{1}{2}(v_{N-1}(t) + v_N(t)) = g_1(t),$$

In matrix form the system ODEs (10.51) becomes

$$\frac{d\mathbf{v}}{dt} = \frac{a}{2h} A \mathbf{v} + \mathbf{F}(t), \tag{10.52}$$

with

$$\mathbf{v} = \begin{pmatrix} v_0 \\ v_1 \\ \vdots \\ v_{N-1} \end{pmatrix}, \quad A = \begin{pmatrix} -2 & 2 & 0 & \cdots & \cdots & 0 \\ -1 & 0 & 1 & \cdots & \cdots & 0 \\ 0 & -1 & 0 & 1 & \ddots & 0 \\ \vdots & \ddots & \ddots & \ddots & \ddots & \\ \vdots & & \ddots & -1 & 0 & 1 \\ 0 & \cdots & & 0 & -1 & -1 \end{pmatrix}, \quad \mathbf{F}(t) = \frac{a}{h} \begin{pmatrix} 0 \\ \vdots \\ 0 \\ g_1(t) \end{pmatrix}.$$

Remark 10.7 *The boundary data function $g_1(t)$ becomes the forcing for the system of ODEs. One can easily check using modern computer software, for various dimensions N, that the matrix A is diagonable and its eigenvalues have negative real part and nonzero imaginary part. The theory of Chapter 6 applies.*

To prove stability of this scheme, and that the semidiscrete problem behaves as the differential problem, we construct an energy estimate. We use the energy

$$E(t) = \frac{h}{2}|v_0|^2 + \|v\|_{1,N-1}^2. \tag{10.53}$$

---

[1] This is equivalent to write a second order accurate evolution for v_0 with a linearly extrapolated value of v_{-1} (i.e., $h^2 D_+ D_- v_0 = 0$).

Then, using Lemma 10.6 and the boundary conditions, we have

$$
\begin{aligned}
\frac{dE}{dt} &= (v_t, v)_{1,N-1} + (v, v_t)_{1,N-1} + \frac{h}{2}(\bar{v}_{0t}v_0 + \bar{v}_0 v_{0t}) \\
&= a\big((D_0 v, v)_{1,N-1} + (v, D_0 v)_{1,N-1}\big) + \frac{a}{2}h\big((D_+\bar{v}_0)v_0 + \bar{v}_0 D_+\bar{v}_0\big) \\
&= \frac{a}{2}(\bar{v}_{N-1}v_N + \bar{v}_N v_{N-1} - \bar{v}_0 v_1 - \bar{v}_1 v_0) + \frac{a}{2}(\bar{v}_1 v_0 + \bar{v}_0 v_1 - 2|v_0|^2) \\
&= a(\bar{v}_{N-1}g_1 + \bar{g}_1 v_{N-1} - |v_{N-1}|^2) - a|v_0|^2 \\
&\le -a(|v_{N-1}|^2 - 2|v_{N-1}||g_1|) \\
&\le a|g_1(t)|^2.
\end{aligned}
\tag{10.54}
$$

Therefore, integrating, we obtain the energy estimate

$$
E(t) \le E(0) + a \int_0^t |g_1(s)|^2 ds.
\tag{10.55}
$$

The semidiscrete system of ODEs behaves as the differential equation (see Lemma 10.1). To integrate in time, one needs to use a method that includes the imaginary axis in its stability region (see Remark 10.7). One can use implicit Euler or the trapezoidal rule. However, for hyperbolic equations such as ours, it is generally more efficient to use an explicit multistep method such as Adams-Bashforth.

Exercise 10.5 *Consider the problem* (10.13)–(10.15) *with* $a = 1$, *and*

$$
f(x) = 0, \quad g_1(t) = \begin{cases} \frac{1}{2}(1 - \cos(2\pi t)), & t \in [0, 1], \\ 0, & t > 1. \end{cases}
$$

(a) Calculate the exact solution $u(x, t)$ *(how smooth is it?)*

(b) Write a computer code to integrate system (10.51) *for* $0 \le t \le 1.5$ *with the two-step Adams-Bashforth method. Use* $h = 1/100$ *in the grid* (10.32), *and time step* $k = ah/4$. *Make plots of the numerical solution and error* $(v_j^{(h)}(t) - u(x_j, t))$ *at* $t = 0.5$, $t = 1.0$ *and* $t = 1.5$.

(c) Make a second run using half the mesh size. At $t = 1$ *compute the integrated precision quotient*

$$
Q = \frac{\|v^{(h)}(t = 1) - u(t = 1)\|_{1,99}}{\|v^{(h/2)}(t = 1) - u(t = 1)\|_{1,199}}.
$$

Do you get a value close to 4?

10.2.5 Wave equation

To finish this chapter we present a second-order approximation to solve the initial value problem for the wave equation on a strip. Instead of approximating the problem (10.17), (10.28), we transform the problem as in Section 10.1.3 so that the problem consists in two scalar one-way wave equations coupled by the boundary conditions. Thus, all we need is a second-order semidiscrete approximation of the problem

(10.23), (10.24), (10.25). We can apply the second-order approximation studied for the one-way wave equation in Section 10.2.4.

Again the grid is given by (10.32). The semidiscretized system is, calling $v_j^i(t)$ the approximation of $w^i(x_j, t)$, $i = 0, 1$,

$$v_j^1(0) = \frac{1}{\sqrt{2}}\left(c\frac{df}{dx}(x_j) + h(x_j)\right), \quad j = 1, 2, \ldots, N-1, \quad v_0^1(0) = v_1^1(0),$$

$$\frac{dv_j^1}{dt} = cD_0 v_j^1, \quad j = 1, 2, \ldots, N-1,$$

$$\frac{dv_0^1}{dt} = cD_+ v_0^1, \quad \frac{1}{2}(v_{N-1}^1(t) + v_N^1(t)) = \frac{R^2}{2}(v_{N-1}^2(t) + v_N^2(t)) + \tilde{g}_1(t),$$

$$v_j^2(0) = \frac{1}{\sqrt{2}}\left(-c\frac{df}{dx}(x_j) + h(x_j)\right), \quad j = 1, 2, \ldots, N-1, \quad v_N^2(0) = v_{N-1}^2(0),$$

$$\frac{dv_j^2}{dt} = -cD_0 v_j^2, \quad j = 1, 2, \ldots, N-1,$$

$$\frac{dv_N^2}{dt} = -cD_- v_N^2, \quad \frac{1}{2}(v_0^2(t) + v_1^2(t)) = \frac{R^1}{2}(v_0^1(t) + v_1^1(t)) + \tilde{g}_0(t).$$

To prove the stability of this approximation, one can use the energy

$$E(t) = \frac{h}{2}|v_0^1(t)|^2 + \|v^1(t)\|_{1,N-1}^2 + \|v^2(t)\|_{1,N-1}^2 + \frac{h}{2}|v_N^2(t)|^2. \tag{10.56}$$

For $R^1 = R^2 = 0$, this scheme is simply two decoupled one-way equations such as those of Section 10.2.4. Therefore, the scheme is stable. When R^i, $i = 0, 1$, are different from zero, one can get an energy estimate of the form

$$E(t) \leq K(t)\left(E(0) + \int_0^t e^{-s}(|\tilde{g}_0(s)|^2 + |\tilde{g}_1(s)|^2)\, ds\right), \tag{10.57}$$

where $K(t)$ is independent of the initial data.

Exercise 10.6 *Prove* (10.57).

To integrate in time, one can use the same methods mentioned in Section 10.2.4. Once the problem is solved in this varables, one can get a second-order approxima-tion of the original function $\varphi(x, t)$ of problem (10.17) since one has second-order approximations for both of its partial derivatives.

CHAPTER 11

NONLINEAR PROBLEMS

In this chapter we discuss nonlinear problems. We are interested in smooth solutions. No general theory for nonlinear differential equations is available. Instead, we ask the following questions. Assume that we know a solution U for a particular set of data. Is the problem still solvable if we make small perturbations of the data? Does the solution depend continuously on the perturbation; that is, do small perturbations in the data generate small changes in the solution?

We can linearize the nonlinear equations around the known solution U and we will see that the properties of this linear system often determine the answer to the questions above.

In practice, one often solves nonlinear problems numerically without having any knowledge as to whether the differential equations have a solution. If the numerical solution is smooth in the sense that it varies slowly with respect to the mesh, we can interpolate the numerical solution. The interpolant solves a nearby problem and the solution of the original problem can be considered a perturbation of the numerically constructed solution. Therefore, the questions above are of interest.

For the sake of simplicity, in this chapter we treat initial value problems for partial differential equations with 1-periodic boundary conditions. Conclusions analogous

Introduction to Numerical Methods for Time Dependent Differential Equations, First Edition. By Heinz-O. Kreiss and Omar E. Ortiz. Copyright © 2014 John Wiley & Sons, Inc.

to those derived in this chapter can be drawn for more general initial boundary value problems.

11.1   Initial value problems for ordinary differential equations

We start with a simple model problem

$$\frac{dy}{dt} = \alpha y + \varepsilon y^2, \quad t \geq 0,$$

$$y(0) = y_0.$$

(11.1)

Here α, ε are real constants with $0 < \varepsilon \ll 1$. We assume also that $y_0 > 0$ and calculate the solution explicitly. Introducing a new variable by

$$y = e^{\alpha t} \tilde{y}$$

gives us

$$\frac{d\tilde{y}}{dt} = \varepsilon e^{\alpha t} \tilde{y}^2,$$

$$\tilde{y}(0) = y_0.$$

Therefore,

$$\frac{1}{y_0} - \frac{1}{\tilde{y}(t)} = \varepsilon \int_0^t e^{\alpha s}\, ds = \varepsilon \psi(t, \alpha),$$

where

$$\psi(t, \alpha) = \begin{cases} \dfrac{1}{\alpha}(e^{\alpha t} - 1) & \text{if } \alpha \neq 0, \\ t & \text{if } \alpha = 0. \end{cases}$$

Solving for $\tilde{y}(t)$ gives us

$$\tilde{y}(t) = \frac{1}{1/y_0 - \varepsilon \psi(t, \alpha)}.$$

(11.2)

There are three different regimes.

- If $\alpha > 0$, the solution blows up and the blow-up time $T = \mathcal{O}(\log(1/\varepsilon\alpha))$. Thus, it does not help very much to decrease ε. T increases only logarithmically.

- If $\alpha = 0$, $T = 1/(y_0\varepsilon)$ and the blow-up time is linear in ε.

- If $\alpha < 0$, there is no blow-up for sufficiently small ε.

The discussion above shows that the sign of α is the dominating factor determining the behavior of the solution. If $\alpha < 0$, we can, for sufficiently small ε, neglect the nonlinear terms. This is also true if $\alpha = 0$, provided that the time we consider is not too long.

The same type of result holds for general equations:

$$\frac{dy}{dt} = \alpha y + \varepsilon F(y, t),$$
$$y(0) = y_0,$$

(11.3)

where $F(y, t)$ denotes the nonlinear term.

We shall now solve a nonlinear equation,

$$\frac{dy}{dt} = f(y, t),$$
$$y(0) = y_0,$$

(11.4)

using the forward Euler method. Let $k > 0$ denote the grid size and $v_n = v(nk)$ the approximation of y on the grid. Euler's forward method can be written as

$$v_{n+1} = v_n + kf(v_n, t_n),$$
$$v_0 = y_0.$$

(11.5)

We calculate v_n in some time interval $0 \le t \le T$ and want to decide whether the numerical solution v has anything to do with the analytic solution y.

It is well known (see, e.g., [5]) that we can interpolate the discrete grid function v by splines such that the resulting interpolant $\varphi = \mathrm{Int}(v)$ belongs to $C^p(0, T)$ and

$$\sum_{j=0}^{p} \left| \frac{d^j \varphi(\cdot)}{dt^j} \right|_{\infty} \le K_p \sum_{j=0}^{p} |D_k^j v|_{k,\infty}.$$

(11.6)

Here

$$\left| \frac{d^j \varphi(\cdot)}{dt^j} \right|_{\infty} = \max_t \left| \frac{d^j \varphi(\cdot)}{dt^j} \right| \quad \text{and} \quad |D_k^j v|_{k,\infty} = \max_l |D_k^j v_l|,$$

where the $D_k^j v_l$ denote the divided differences of order j. We can choose p arbitrarily. The constants K_p increase with p and are, for $p \le 10$, of moderate size. The estimate (11.6) is similar to the one for the periodic case that we derived in Theorem 9.2.

We want to show that $\varphi(t)$ solves a nearby differential equation. We need

Lemma 11.1 *Consider a time interval* $0 \le t \le T$ *which we cover by a grid* $t_n = nk$, $n = 1, 2, 3, \ldots, N$; $Nk = T$. *Let* $F \in C^1$ *be a function with*

$$\max_{0 \le n \le N} |F(t_n)| = \delta.$$

Then

$$|F(\cdot)|_{\infty} \le \delta + k \left| \frac{dF}{dt}(\cdot) \right|_{\infty}, \quad |F(\cdot)|_{\infty} = \max_{0 \le t \le T} |F(t)|.$$

Proof: For $t \in (0, T)$, there are grid points t_n, t_{n+1} such that

$$t_n \le t \le t_{n+1}.$$

Therefore,

$$F(t) = F(t_n) + \int_{t_n}^{t} F_t(s)\, ds,$$

and the lemma follows.   ∎

Without proof we state the following generalization.

Lemma 11.2 *If F in Lemma 11.1 belongs to C^p, there is a constant C_p such that*

$$|F(t)|_\infty \le \delta + C_p k^p \left| \frac{d^p F(\cdot)}{dt^p} \right|_\infty.$$

We assume now that the interpolant φ belongs to C^2. Then $d\varphi/dt - f(\varphi, t) \in C^1$ and we have a bound for

$$\frac{d}{dt}\left(\frac{d\varphi}{dt} - f(\varphi, t) \right) = \frac{d^2\varphi}{dt^2} - \frac{\partial f}{\partial \varphi}\frac{d\varphi}{dt} - \frac{\partial f}{\partial t}.$$

We write, for arbitrary t,

$$\frac{d\varphi}{dt} = f(\varphi(t), t) + kR(t) \tag{11.7}$$

and bound $kR(t)$ using Lemma 11.1. For every grid point, by Taylor's theorem,

$$\left| \frac{d\varphi}{dt}(t_n) - f(\varphi(t_n), t_n) \right| = \left| \frac{d\varphi}{dt}(t_n) - \frac{\varphi(t_{n+1}) - \varphi(t_n)}{k} \right|$$

$$\le k \left| \frac{d^2\varphi}{dt^2}(\cdot) \right|_\infty.$$

Therefore, by Lemma 11.1, (11.7) holds with

$$|R(t)| \le \left| \frac{d^2\varphi}{dt^2}(\cdot) \right|_\infty + \left| \frac{d}{dt}\left(\frac{d\varphi}{dt} - f(\varphi, t) \right) \right|_\infty.$$

$R(t)$ is, essentially, the truncation error evaluated at the interpolant. If $\varphi \in C^p$, then $R(t) \in C^{p-2}$ and we have bounds for the derivative of R in terms of the divided differences of the numerical solution. Thus, if the numerical solution "looks smooth" which means that the divided differences are bounded, the interpolant solves a nearby differential equation given by (11.7).

This is as close as numerical methods can get us to the true solution. If we want to know how close we are to the true solution $y(t)$, we have to use perturbation theory. We make the change of variables $y(t) = \varphi(t) + k\tilde{y}(t)$. By Taylor expansion, (11.4) becomes

$$\frac{d\varphi}{dt} + k\frac{d\tilde{y}}{dt} = f(\varphi + k\tilde{y}, t) = f(\varphi, t) + \frac{\partial f}{\partial \varphi}(\varphi, t)\tilde{y} + \mathcal{O}(k^2 \tilde{y}^2),$$

and the problem for $\tilde{y}$, equivalent to (11.4), is

$$\frac{d\tilde{y}}{dt} = \frac{\partial f}{\partial \varphi}(\varphi, t)\tilde{y} + kg(\tilde{y}, t) - R(t),$$

$$\tilde{y}(0) = 0.$$

(11.8)

Here $g(\tilde{y}, t)$ is quadratic in $\tilde{y}$. As in the model problem, the behavior of the linearized problem

$$\frac{d\tilde{\tilde{y}}}{dt} = \frac{\partial f}{\partial \varphi}(\varphi, t)\tilde{\tilde{y}} - R(t),$$

$$\tilde{\tilde{y}}(0) = 0,$$

(11.9)

tells us how long $\tilde{y}(t)$ stays bounded. If the solution operator of (11.9) decays exponentially, then for sufficiently small k, $\tilde{y}(t)$ stays bounded for all times. On the other hand, if the solution operator grows exponentially, the blow-up can occur at $t = \mathcal{O}\big(\log(1/k)\big)$.

For complicated problems, one has no analytical knowledge of the behavior of the solution operator. Therefore, one relies on numerical perturbation calculations.

Instead of the Euler method, we could have used a higher-order method such as the fourth-order Runge-Kutta method. In that case the forcing in equation (11.7) would have been of order k^4.

The procedure can also be used for partial differential equations, but the estimates of the derivatives of the interpolant become rather complicated.

11.2   Existence theorems for nonlinear partial differential equations

Consider the Cauchy problem for a quasilinear first-order partial differential equation

$$\frac{\partial \tilde{u}}{\partial t} = P\Big(x, t, \tilde{u}, \frac{\partial}{\partial x}\Big)\tilde{u} + F_1(x, t),$$

$$\tilde{u}(x, 0) = f_1(x),$$

(11.10)

where

$$P\Big(x, t, \tilde{u}, \frac{\partial}{\partial x}\Big)\tilde{u} = a(x, t, \tilde{u})\frac{\partial}{\partial x}\tilde{u} + b(x, t, \tilde{u})\tilde{u}.$$

Even if we assume that all coefficients and data are smooth functions of all variables, no global existence theory is available. The only general results are of local character which can be phrased in the following way. Assume that we know that a nearby problem,

$$\frac{\partial U}{\partial t} = P\Big(x, t, U, \frac{\partial}{\partial x}\Big)U + F_1(x, t) - \varepsilon F(x, t),$$

$$U(x, 0) = f_1(x) - \varepsilon f(x),$$

(11.11)

has a smooth solution in some time interval $0 \le t \le T$. Here $\varepsilon > 0$ is a small constant. Can we infer that for sufficiently small ε, the original problem (11.10) has a solution in the same time interval and that $|U - u| = \mathcal{O}(\varepsilon)$? Using the arguments of Section 11.1, U can often, at least in principle, be obtained by interpolating a numerical solution of the problem.

The change of variables

$$\tilde{u} = U + \varepsilon u$$

leads to the system

$$\frac{\partial u}{\partial t} = P_0\left(x, t, \frac{\partial}{\partial x}\right) u + \varepsilon P_1\left(x, t, u, \frac{\partial}{\partial x}\right) u + F(x, t),$$
$$u(x, 0) = f(x).$$
(11.12)

Here P_0 denotes the linear operator that one obtains by linearizing $P(x, t, \tilde{u}, \partial/\partial x)\tilde{u}$ around U. A natural assumption is that the linear problem

$$\frac{\partial w}{\partial t} = P_0\left(x, t, \frac{\partial}{\partial x}\right) w + F(x, t),$$
$$w(x, t) = f(x).$$
(11.13)

is a well-posed problem. We shall now give arguments that under reasonable assumptions, this assumption guarantees that (11.12) also as a solution, provided that ε is sufficiently small. We start with an example and consider

$$\frac{\partial u}{\partial t} = \alpha u + \varepsilon u \frac{\partial u}{\partial x} + F(x, t),$$
$$u(x, 0) = f(x).$$
(11.14)

Here α, ε are real constants and $f, F \in C^\infty$ are real smooth functions which are 2π-periodic in x. We are interested in real solutions that are also 2π-periodic in x.

The most important tools to derive existence theorems are a priori estimates, that is, we assume that there is a smooth solution and derive estimates of u and its derivatives in terms of f and F and their derivatives. Once one has obtained these estimates, existence follows. Here we derive only a priori estimates. We refer to the literature (e.g., Chapter 4 of [7]) for a discussion of how to use them for existence theorems.

As before, $u(x, t)$, $\|u\|^2 = (u, u)$ denote the L_2-scalar product and norm, here with respect to the interval $0 \le x \le 2\pi$. Multiplying (11.14) with u and integrating gives

$$\frac{1}{2}\frac{d}{dt}(u, u) = \alpha\|u\|^2 + \varepsilon\left(u, u\frac{\partial u}{\partial x}\right) + (u, F).$$
(11.15)

Integration by parts implies that

$$\left(u, u\frac{\partial u}{\partial x}\right) = \left(u^2, \frac{\partial u}{\partial x}\right) = -2\left(u\frac{\partial u}{\partial x}, u\right) = -2\left(u, u\frac{\partial u}{\partial x}\right),$$

that is,

$$\left(u, u\frac{\partial u}{\partial x}\right) = 0.$$

Therefore, (11.15) gives us

$$\frac{1}{2}\frac{d}{dt}\|u\|^2 \leq \alpha\|u\|^2 + \|u\|\,\|F\|,$$

that is,

$$\frac{d}{dt}\|u\| \leq \alpha\|u\| + \|F\|, \quad \|u(\cdot,0)\| = \|f\|.$$

Therefore,

$$\|u(\cdot,t)\| \leq e^{\alpha t}\|f\| + \int_0^t e^{\alpha(t-s)}\|F(\cdot,s)\|\,ds$$
$$\leq e^{\alpha t}\|f\| + \max_{0\leq s\leq t}\|F(\cdot,s)\|\,\psi(\alpha,t), \tag{11.16}$$

where

$$\psi(\alpha,t) = \begin{cases} (e^{\alpha t}-1)/\alpha & \text{for } \alpha \neq 0, \\ t & \text{for } \alpha = 0. \end{cases}$$

As in the ODE case, the sign of α determines whether the solution grows or stays bounded.

To obtain bounds for $v = du/dx$ we differentiate (11.14) with respect to x and obtain

$$\frac{\partial v}{\partial t} = \alpha v + \varepsilon u\frac{\partial v}{\partial x} + \varepsilon v^2 + \frac{\partial F}{\partial x},$$
$$v(x,0) = \frac{df}{dx}. \tag{11.17}$$

Therefore,

$$\frac{1}{2}\frac{d}{dt}\|v\|^2 = \alpha\|v\|^2 + \varepsilon\left(v, u\frac{\partial v}{\partial x}\right) + \varepsilon(v,v^2) + \left(v, \frac{\partial F}{\partial x}\right).$$

Since

$$\left(v, u\frac{\partial v}{\partial x}\right) = -\left(\frac{\partial v}{\partial x}u, v\right) - (v,v^2), \quad (v,v^2) \leq |v|_\infty\|v\|^2,$$

it follows that

$$\left(v, u\frac{\partial v}{\partial x}\right) = -\frac{1}{2}(v,v^2)$$

and

$$\frac{1}{2}\frac{d}{dt}\|v\|^2 \leq \alpha\|v\|^2 + \frac{|\varepsilon|}{2}|v|_\infty\|v\|^2 + \|v\|\left\|\frac{\partial F}{\partial x}\right\|,$$

that is,

$$\frac{d}{dt}\|v\| \leq \alpha\|v\| + \frac{|\varepsilon|}{2}|v|_\infty\|v\| + \left\|\frac{\partial F}{\partial x}\right\|,$$
$$\|v(\cdot,0)\| = \left\|\frac{df}{dx}\right\|. \tag{11.18}$$

Since $|v|_\infty$ cannot be estimated in terms of $\|v\|$, we cannot use (11.18) directly to estimate v.

Differentiating (11.17) gives an equation for $w = \partial^2 u/\partial x^2$:

$$\frac{\partial w}{\partial t} = \alpha w + \varepsilon u \frac{\partial w}{\partial x} + 3\varepsilon v w + \frac{\partial^2 F}{\partial x^2}.$$

Therefore,

$$\frac{1}{2}\frac{d}{dt}\|w\|^2 = \alpha\|w\|^2 + \varepsilon\left(w, u\frac{\partial w}{\partial x}\right) + 3\varepsilon(w, vw) + \left(w, \frac{\partial^2 F}{\partial x^2}\right).$$

Since

$$\left(w, u\frac{\partial w}{\partial x}\right) = -\frac{1}{2}(w, vw),$$

we obtain

$$\frac{d}{dt}\|w\| \leq \alpha\|w\| + \frac{5}{2}|\varepsilon||v|_\infty\|w\| + \left\|\frac{\partial^2 F}{\partial x^2}\right\|$$

$$\leq \alpha\|w\| + \frac{5}{4}|\varepsilon|\left(|v|_\infty^2 + \|w\|^2\right) + \left\|\frac{\partial^2 F}{\partial x^2}\right\|. \tag{11.19}$$

We now derive a Sobolev inequality to estimate $|v|_\infty$ in terms of $\|v\|$, $\|w\|$. Let x_1, x_0 be two points with

$$|v|_\infty = |v(x_1)|, \qquad \min_{0\leq x\leq 2\pi}|v(x)| = v(x_0);$$

then

$$\int_{x_0}^{x_1} vw\,dx = \int_{x_0}^{x_1}\frac{\partial u}{\partial x}\frac{\partial^2 u}{\partial x^2}\,dx = \frac{\partial u}{\partial x}\Big|_{x_0}^{x_1} - \int_{x_0}^{x_1}\frac{\partial u}{\partial x}\frac{\partial^2 u}{\partial x^2}\,dx.$$

Since

$$\min_{0\leq x\leq 2\pi}|v(x)|^2 \leq \frac{1}{2\pi}\|v\|^2, \tag{11.20}$$

we obtain

$$|v|_\infty^2 \leq \frac{1}{2\pi}\|v\|^2 + 2\|v\|\|w\|, \tag{11.21}$$

and then (11.19) becomes

$$\frac{d}{dt}\|w\| \leq \alpha\|w\| + \frac{5}{4}|\varepsilon|\left(\frac{1}{2\pi}\|v\|^2 + 2\|v\|\|w\| + \|w\|^2\right) + \left\|\frac{\partial^2 F}{\partial x^2}\right\|,$$

$$\|w(\cdot, 0)\| = \left\|\frac{\partial^2 f}{\partial x^2}\right\|. \tag{11.22}$$

Together (11.18) and (11.22) represent a closed system of differential inequalities for $\|v\|$ and $\|w\|$. If we replace the inequality sign by an equality sign, we obtain a system of differential equations that maximizes the inequalities. This system is of the same form as the model problem in Section 11.1. The blow-up time, if any, depends on the sign of α and the size of $|\varepsilon|$.

There are no difficulties in estimating higher derivatives. They exist as long as $\partial u/\partial x$, $\partial^2 u/\partial x^2$ stay bounded.

The estimates above can be generalized to rather general mixed hyperbolic-parabolic systems. Consider, for example, the Cauchy problem quasilinear first-order system

$$\frac{\partial \mathbf{u}}{\partial t} = P_0\left(\mathbf{x}, t, \frac{\partial}{\partial x}\right)\mathbf{u} + \varepsilon P_1\left(\mathbf{x}, t, \mathbf{u}, \frac{\partial}{\partial x}\right) + \mathbf{F},$$

$$\mathbf{u}(x, 0) = \mathbf{f}. \tag{11.23}$$

Here $\mathbf{F} = \mathbf{F}(\mathbf{x}, t)$, $\mathbf{f} = \mathbf{f}(\mathbf{x}, t)$, $\mathbf{u} = \mathbf{u}(\mathbf{x}, t)$ are vector-valued functions with n components depending on $\mathbf{x} = (x_1, x_2, \ldots, x_s) \in \mathbb{R}^s$ and t.

$$P_0 = \sum_{j=0}^{s} A_j^{(0)}(\mathbf{x}, t)\frac{\partial}{\partial x_j} + B^{(0)}(\mathbf{x}, t),$$

$$P_1 = \sum_{j=1}^{s} A_j^{(1)}(\mathbf{x}, t, \mathbf{u})\frac{\partial}{\partial x_j} + B^{(1)}(\mathbf{x}, t, \mathbf{u})$$

are first-order operators with symmetric matrix coefficients that depend smoothly on all variables.

Integration by parts shows that there is a constant α such that

$$(w, P_0 w) \leq \alpha\|w\|^2 \quad \text{for all smooth } w.$$

We have

Theorem 11.3 *The system* (11.23) *has a smooth solution in some time interval* $0 \leq t \leq T$; T *depends on the sign of* α *and on* $\mathbf{f}$, $\mathbf{F}$, *and* ε:
If $\alpha < 0$, *then* $T = \infty$ *if* ε *is sufficiently small.*
If $\alpha = 0$, *then* T *is of the order* $\mathcal{O}(1/\varepsilon)$.
If $\alpha > 0$, *then* T *is of the order* $\mathcal{O}(\log(1/\varepsilon))$.

11.3   Nonlinear example: Burgers' equation

We consider as an example the 1-periodic initial value problem for the viscous Burgers' equation

$$\frac{\partial u}{\partial t} = -u\frac{\partial u}{\partial x} + \nu\frac{\partial^2 u}{\partial x^2}, \quad u(x+1, t) = u(x, t), \quad \nu \geq 0,$$

$$u(x, 0) = f(x). \tag{11.24}$$

Burgers' equation is a particular case in which a rather complete analytical study can be carried out. Assume that the initial data is C^∞ smooth. It can be shown that if $\nu = 0$—the inviscid Burgers' equation—a solution exists and is C^∞ smooth only during a finite time because a shock may form. The existence time depends on the

Table 11.1 Results of (11.26) for three runs with different viscosities.

ν	h	$t_{\max}$	$|D_0 v(t_{\max})|$
0.1	$1/125$	0.033184	6.532661
0.01	$1/250$	0.273336	40.04797
0.001	$1/500$	0.254222	489.8528

initial data. On the other hand, it can be shown that for positive ν the solution exists for all times. In the latter case, for well chosen initial data, a shock tends to form, but the second derivative on the right-hand side in (11.24) dissipates energy and the solution remains C^∞ smooth for all times. A detailed analysis of this equation, including energy estimates, existence, and smoothness can be found in Chapter 4 of reference [7].

Intuitively, the equation looks similar to the one-way wave equation we studied before, but now the propagation speed is the solution u itself. In a region where $u > 0$, the solution resembles a wave that moves to the right, whereas in a region where $u < 0$, the solution resembles a wave that moves to the left. If we choose

$$f(x) = \sin(2\pi x), \tag{11.25}$$

there is initially a positive peak at $x = \frac{1}{4}$ and a negative peak at $x = \frac{3}{4}$. With this initial data, Burgers' equation will try to form a shock (diverging $|\partial u/\partial x|$) at the middle of the domain, $x = \frac{1}{2}$, in finite time ($t \approx 0.25$).

Here we want to investigate the problem (11.24), (11.25) numerically. We approximate the problem on the grid $x_j = hj$, $h = 1/N$, $j \in \mathbb{Z}$. We call $v_j^{(h)}(t)$ the approximation of $u(x_j, t)$ on the grid. We approximate the first derivative by D_0 and the second derivative by $D_+ D_-$ and integrate in time with Euler's method using time step $k = h^2/2$, so that the overall method becomes second-order accurate in h. Thus, we solve, with $v_{j+N}^{(h)}(t) = v_j^{(h)}(t)$,

$$v_j^{(h)}(t + k) = v_j^{(h)} D_0 v_j^{(h)} + \nu D_+ D_- v_j^{(h)}, \quad j = 0, 1, \ldots, N - 1,$$
$$v_j^{(h)}(0) = \sin(2\pi x_j). \tag{11.26}$$

We compute the solution $v_j^{(h)}$ for three positive values of ν and look for the time $t_{\max}$ at which $D_0 v_j^{(h)}$ becomes maximum in absolute value (i.e., when the solution is closest to form a shock). For smaller values of ν we use smaller values of h, so that our code resolves well the high values of derivatives that appear. Once we find the value of $t_{\max}$ we compute a second and third solution using $h/2$ and $h/4$ up to that time and evaluate the precision quotient

$$Q_j(t_{\max}) = \frac{v_j^{(h)}(t_{\max}) - v_j^{(h/2)}(t_{\max})}{v_j^{(h/2)}(t_{\max}) - v_j^{(h/4)}(t_{\max})} \tag{11.27}$$

to check that our code is running with correct precision order ($Q_j \simeq 2^2 = 4$), and that the values of h are small enough. We show the results in the Table 11.1 and Figure 11.1.

It is interesting to notice in the plots of the solution that the smaller the value of ν, the less dissipation there is (the initial amplitude is better preserved) and the closer the solution is to develop a shock (see Figure 11.1). As time increases after $t_{\max}$ the energy dissipates and the amplitude decreases.

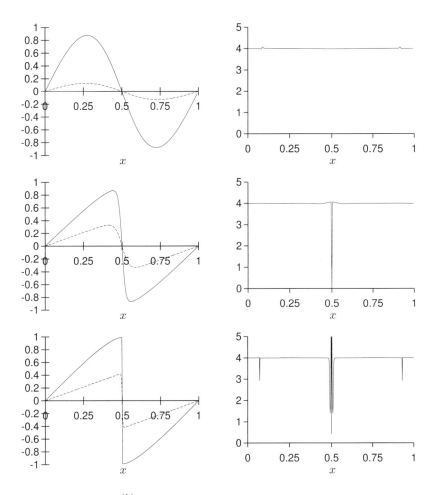

Figure 11.1   Plots of $v_j^{(h)}$ and Q_j for solutions corresponding to table 11.1. The upper left plot is v_j for $\nu = 0.1$ at $t = t_{max}$ (solid line) and $t = 0.5$ (dashed line). The upper right plot is $Q(t_{max})$. The middle left plot is the solution v_j for $\nu = 0.01$ at $t = t_{max}$ (solid line) and $t = 1.0$ (dashed line). The middle right plot is $Q(t_{max})$. The lower plots are like the middle ones but for $\nu = 0.001$.

Exercise 11.1 *Write a computer code to implement* (11.26) *and reproduce the values in Table 11.1 and the plots in Figure 11.1.*

APPENDIX A

AUXILIARY MATERIAL

A.1 Some useful Taylor series

We collect here some useful Taylor series used throughout the text. z denotes a complex variable, and the expression on the right denotes the convergence region.

1. Finite geometric series:

$$\frac{1 - z^{N+1}}{1 - z} = 1 + z + z^2 + \cdots + z^N = \sum_{j=0}^{N} z^j, \quad z \neq 1. \quad \text{(A.1)}$$

2. Geometric series:

$$\frac{1}{1 - z} = 1 + z + z^2 + z^3 + \cdots = \sum_{j=0}^{\infty} z^j, \quad |z| < 1. \quad \text{(A.2)}$$

3. Any branch of the logarithm has the expansion:

$$\log(1 + z) = z - \frac{z^2}{2} + \frac{z^3}{3} - \frac{z^4}{4} + \cdots = \sum_{j=1}^{\infty} \frac{(-1)^{j+1}}{j} z^j, \quad |z| < 1. \quad \text{(A.3)}$$

Introduction to Numerical Methods for Time Dependent Differential Equations, First Edition.
By Heinz-O. Kreiss and Omar E. Ortiz. Copyright © 2014 John Wiley & Sons, Inc.

A.2   "$\mathcal{O}$" notation

Definition A.1 *Let $F(x)$ be a real function. We say that "$F(x)$ is of order $f(x)$ as x tends to a" and write $F(x) = \mathcal{O}(f(x))$ as $x \to a$, if and only if there exist two constants $C > 0$ and $\delta > 0$ such that $|F(x)| \leq C\,|f(x)|$ when $|x - a| < \delta$.*

For example, if $F(x)$ is $n+1$ times differentiable around $x = a$, then by Taylor's theorem,

$$F(x) = F(a) + \frac{dF}{dx}(a)(x-a) + \frac{1}{2}\frac{dF}{dx}(a)(x-a)^2 + \cdots + \frac{1}{n!}\frac{d^n F}{dx^n}(a)(x-a)^n + R_n(x),$$

where the remainder term satisfies $R_n(x) = \mathcal{O}((x-a)^{n+1})$. Here, and many times when it is clear by context, we omit "as $x \to a$."

A.3   Solution expansion

In this section we give a derivation of the *solution expansion*: a fundamental expansion in powers of k that relates the numerical approximation to the exact solution of an initial value problem. We derive this expansion for the particular case of the explicit Euler method applied to a linear equation; we then generalize the result, without proof, to other methods.

Consider the scalar initial value problem

$$\frac{dy}{dt} = a(t)y + F(t),$$
$$y(0) = y_0,$$

(A.4)

where $a(t)$ and $F(t)$ are smooth functions of t.[1] By the existence and uniqueness theorem for ordinary differential equations, this problem has a unique smooth solution $y(t)$.

Now, approximate (A.4) by the explicit Euler method:

$$v(t + k) = v(t) + ka(t)v(t) + kF(t),$$
$$v(0) = y_0.$$

(A.5)

Here $a(t)$ and $F(t)$ are smooth functions and so is the exact solution $y(t)$. We want to derive an asymptotic expansion, in the variable k, of the error of the approximation $v(t, k) - y(t)$. Use of the notation $v(t, k)$ instead of our usual notation $v(t)$ is to emphasize the dependence of the approximation on the step size k.

The explicit Euler approximation (A.5) is globally accurate of order $\mathcal{O}(k)$; therefore, we write $v(t, k) = y(t) + \mathcal{O}(k)$ within the time interval of interest. To make

---

[1]Throughout this appendix, "smooth" means C^∞-smooth. This smoothness condition can be relaxed depending on how large the order is that one wants to make explicit in the expansion, but we do not go into the details here.

explicit the form of the $\mathcal{O}(k)$ term, we proceed as follows. Substituting the solution $y(t)$ of the differential equation into the difference approximation and using Taylor expansion gives us k times the truncation error,

$$
y(t+k) - y(t) - ka(t)y(t) - kF(t) = y(t) + k\frac{dy}{dt}(t)
$$
$$
+ \frac{k^2}{2}\frac{d^2y}{dt^2}(t) + \frac{k^3}{6}\frac{d^3y}{dt^3}(t) + \mathcal{O}(k^4) - y(t) - ka(t)y(t) - kF(t)
$$
$$
= \frac{k^2}{2}\frac{d^2y}{dt^2}(t) + \frac{k^3}{6}\frac{d^3y}{dt^3}(t) + \mathcal{O}(k^4). \quad \text{(A.6)}
$$

We subtract equation (A.6) from (A.5) and obtain for the divided error, $e(t, k) = (v(t, k) - y(t))/k$,

$$
e(t+k,k) - e(t,k) - ka(t)e(t) + \frac{k}{2}\frac{d^2y}{dt^2}(t) = \frac{k^2}{6}\frac{d^3y}{dt^3}(t) + \mathcal{O}(k^3) \quad \text{(A.7)}
$$
$$
e(0, k) = 0.
$$

Here we think d^2y/dt^2 and d^3y/dt^3 as given functions. To first order in k [i.e., neglecting $\mathcal{O}(k^2)$ terms], (A.7) is simply the explicit Euler method applied to the initial value problem

$$
\frac{d\varphi_1}{dt} = a(t)\varphi_1 - \frac{1}{2}\frac{d^2y}{dt^2}(t), \quad \text{(A.8)}
$$
$$
\varphi_1(t) = 0,
$$

which, as $y(t)$ is smooth, has a unique, smooth exact solution $\varphi_1(t)$. We emphasize here that neither the initial value problem (A.8) nor its solution $\varphi_1(t)$ depend on k. Now, as (A.7) is a difference approximation of (A.8) which is globally accurate of order $\mathcal{O}(k)$, we write its solution $e(t, k)$ as $e(t, k) = \varphi_1(t) + \mathcal{O}(k)$. That is, the zero-order term in the asymptotic expansion of $e(t, k)$ is $\varphi_1(t)$. To make explicit the $\mathcal{O}(k)$ term in the expansion of $e(t, k)$, we repeat the process. By Taylor expansion and using (A.8), we have

$$
\varphi_1(t+k) - \varphi_1(t) - ka(t)\varphi_1(t) + \frac{k}{2}\frac{d^2y}{dt^2}(t)
$$
$$
= \frac{k^2}{2}\frac{d^2\varphi_1}{dt^2}(t) + \frac{k^3}{6}\frac{d^3\varphi_1}{dt^3}(t) + \mathcal{O}(k^4). \quad \text{(A.9)}
$$

Subtracting (A.9) from (A.7) gives us for the divided error $e_1(t) = (e(t) - \varphi_1(t))/k$,

$$
e_1(t+k) = e_1(t) + ka(t)e_1(t) + \frac{k}{6}\left(\frac{d^3y}{dt^3} - 3\frac{d^3\varphi_1}{dt^3}\right) + \mathcal{O}(k^2), \quad \text{(A.10)}
$$
$$
e_1(0) = 0.
$$

As before, to first order in k, (A.10) is the explicit Euler approximation of the problem

$$\frac{d\varphi_2}{dt} = a(t)\varphi_2 - \frac{1}{6}\left(\frac{d^3y}{dt^3} - 3\frac{d^2\varphi_1}{dt^2}\right),$$
$$\varphi_2(0) = 0. \tag{A.11}$$

As $y(t)$ and $\varphi_1(t)$ are smooth, the problem (A.11) has a unique solution $\varphi_2(t)$ which is independent of k. Now, as (A.10) is a difference approximation of (A.11), which is globally accurate of order $\mathcal{O}(k)$, we write its solution $e(t,k)$ as $e(t,k) = \varphi_1(t) + \mathcal{O}(k)$. In this way we get

$$
\begin{aligned}
v(t,k) &= y(t) + ke(t,k) \\
&= y(t) + k\varphi_1(t) + k^2 e_1(t,k) \\
&= v(t) + k\varphi_1(t) + k^2\varphi_2(t) + \mathcal{O}(k^3).
\end{aligned}
$$

Of course, one could expand the truncation error (A.6) to higher order in k. In that case we could also repeat the process above to higher order. To summarize, by the procedure above one can show the following representation of the solution $v(t,k)$ of the explicit Euler approximation to the initial value problem (A.4).

Theorem A.2 *Given any positive integer N, the solution $v(t,k)$ of (A.2) can be written as*

$$v(t,k) = y(t) + k\varphi_1(t) + k^2\varphi_2(t) + \cdots + k^N\varphi_N(t) + \mathcal{O}(k^{N+1}). \tag{A.12}$$

Here the functions $\varphi_j(t)$ are solutions of linear inhomogeneous differential equations, which do not depend on k.

By the procedure used above, we can prove a result valid for a general one-step approximation to an initial value problem. Consider an initial value problem

$$\frac{dy}{dt} = f(y,t), \quad y(t=0) = y_0, \tag{A.13}$$

where $f(y,t)$ is assumed to be a smooth function; thus, there exists a unique exact solution $y(t)$ and a $T > 0$ such that $y(t)$ is smooth for $t \in [0,T]$.

We approximate (A.13) by a one-step numerical method

$$
\begin{aligned}
v(t+k,k) &= v(t,k) + k\Phi(v,t,k), \\
v_0 &= y_0,
\end{aligned}
\tag{A.14}
$$

locally accurate of order $\mathcal{O}(k^{p+1})$. As before, we write $v(t,k)$ to emphasize the dependence of v on the step size k.

We state without proof the following result.

Theorem A.3 *Given a positive integer N, there exists a positive constant K_N and a positive time $T_N \leq T$ such that the solution $v(k, t)$ of (A.14) admits the expansion*

$$v(t, k) = y(t) + k^p \varphi_p(t) + k^{p+1} \varphi_{p+1}(t) + \cdots + k^{p+N} \varphi_{p+N}(t) + \mathcal{O}(k^{p+N+1})$$
(A.15)

for $k \leq K_N$ and $t \in [0, T_N]$. Here the functions $\varphi_j(t)$ are solutions of nonlinear equations that do not depend on k.

APPENDIX B

SOLUTIONS TO EXERCISES

SOLUTIONS FOR CHAPTER 1

1.3 Duhamel's principle and integration by parts give the answer. One gets $\tilde{y}_0 = y_0$ in the resonance case and $\tilde{y}_0 = y_0 - (-1)^n P_n^{(n)}(0)/(\mu - \lambda)^{n+1}$ in the nonresonance case.

1.4

$$y(t) = \frac{37}{32}e^{-2t} + \frac{1}{32}\left(\sin(2t)(3 + 4t^2) + \cos(2t)(-5 + 4t - 4t^2)\right).$$

1.5 (a)
$$S(t_2, t_1) = \ln\left(\frac{1 + t_2}{1 + t_1}\right), \quad y(t) = 1 + 2t - \ln(1 + t).$$

(b)
$$S(t_2, t_1) = e^{t_1^2} - e^{t_2^2}, \quad y(t) = e^{-t^2} - e^{-2t^2}.$$

(c)
$$S(t_2, t_1) = e^{\sin(t_2) - \sin(t_1)}, \quad y(t) = e^{-\sin(t)}(1 - \sin(t)) + 1.$$

Introduction to Numerical Methods for Time Dependent Differential Equations, First Edition. By Heinz-O. Kreiss and Omar E. Ortiz. Copyright © 2014 John Wiley & Sons, Inc.

1.7 Assume that $y(t)$ is a smooth solution that blowsup when $t \to T_0 < \infty$. By the mean value theorem, for each t such that $0 < t < T_0$, there exists a time $\tilde{t}$, with $0 \le \tilde{t} \le t$, such that

$$\frac{dy}{dt}(\tilde{t}) = \frac{y(t) - y_0}{t}.$$

Taking the limit $t \to T_0^-$, the right-hand side diverges and then dy/dt diverges at a time $\tilde{T}_0 \le T_0$. But $\tilde{T}_0$ cannot be strictly smaller than T_0 because the differential equation forbids dy/dt to diverge when y is finite.

1.9 No. By exercise 1.7 for a smooth solution to blowup in finite time, its derivative needs to diverge too, but this is forbidden by the equation since $\sin(y)$ is a bounded function. More explicitly, the differential equation implies that

$$-1 \le -|\sin(y)| \le \frac{dy}{dt} \le |\sin(y)| \le 1.$$

Integrating, we have

$$y_0 - t \le y(t) \le y_0 + t,$$

showing that $y(t)$ cannot diverge in finite time.

SOLUTIONS FOR CHAPTER 2

2.1 (a) The analytic solution is

$$y(t) = \left(1 + \frac{2\pi}{1 + 4\pi^2}\right)e^{-t} + \frac{1}{1 + 4\pi^2}\left(\sin(2\pi t) - 2\pi \cos(2\pi t)\right).$$

The solution shows an exponentially decaying transient term plus an oscillatory term. The plot of the solution for $t \in [0, 2]$ is

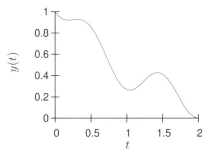

(b) and (c) The plots of the numerical solution and error, for $k = 0.1$, are

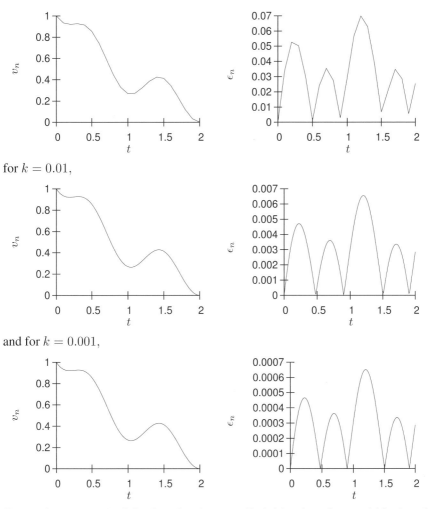

for $k = 0.01$,

and for $k = 0.001$,

Comparing the graphs, it is clear that the error diminishes by a factor of 10 when the timestep k is diminished by the same factor. This is correct since the global error for the Euler method is of order $\mathcal{O}(k)$.

2.2 (a) $y(t) = 3e^t - 2 - 2t - t^2$.

(b) The plot of $\epsilon(k)$ is

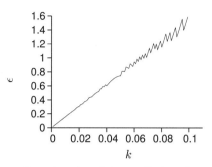

Clearly, $\epsilon(k)$ behaves linearly when k is small; therefore, there is a linear bound for $\epsilon(k)$ but not a quadratic bound.

2.3 By the solution expansion (A.15), we have, for example for $\tilde{Q}$,

$$
\begin{aligned}
\tilde{Q}(t) &= \frac{v^{(1)}(t,k) - v^{(2)}(t,k/2)}{v^{(2)}(t,k/2) - v^{(3)}(t,k/4)} \\
&= \frac{(k^p - (k/2)^p)\varphi_p(t) + \mathcal{O}(k^{p+1})}{((k/2)^p - (k/4)^p)\varphi_p(t) + \mathcal{O}(k^{p+1})} \\
&= 2^p \frac{(k^p - (k/2)^p)\varphi_p(t) + \mathcal{O}(k^{p+1})}{(k^p - (k/2)^p)\varphi_p(t) + \mathcal{O}(k^{p+1})} \\
&= 2^p(1 + \mathcal{O}(k)) \\
&= 2^p + \mathcal{O}(k).
\end{aligned}
$$

2.4 Plots of $Q(t)$, using $k = 0.01$ on the left and using $k = 0.001$ on the right:

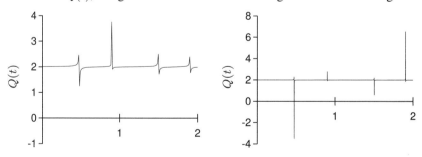

2.5 By (2.39)–(2.41),

$$
w(t,k) = y(t) + \frac{17}{192}k^3\varphi_3(t) + \mathcal{O}(k^4).
$$

Therefore,

$$
Q(t) := \frac{w(t,k/2) - w(t,k)}{w(t,k/4) - w(t,k/2)} = 2^3 + \mathcal{O}(k).
$$

SOLUTIONS FOR CHAPTER 3

3.2 According to Theorem A.3, the solutions $v^{(1)}(nk)$ and $v^{(2)}\big((2n)(k/2)\big)$ satisfy the expansions

$$
\begin{aligned}
v^{(1)}(nk) - y(nk) &= k^p \, \varphi_p(nk) + \mathcal{O}(k^{p+1}), \\
v^{(2)}\big((2n)(k/2)\big) - y(nk) &= \frac{k^p}{2^p} \, \varphi_p(nk) + \mathcal{O}(k^{p+1}),
\end{aligned}
\tag{A.1}
$$

and therefore

$$
v^{(1)}(nk) - v^{(2)}\big((2n)(k/2)\big) = \left(1 - \frac{1}{2^p}\right) k^p \, \varphi_p(nk) + \mathcal{O}(k^{p+1}).
\tag{A.2}
$$

Thus, if $\big|v^{(2)}\big((2n)(k/2)\big) - v^{(1)}(nk)\big| \le (2^p - 1)E$, we have by (A.2), neglecting terms of order k^{p+1},

$$
\left(1 - \frac{1}{2^p}\right) k^p |\varphi_p(nk)| \le (2^p - 1)E,
$$

and therefore

$$
\frac{1 - 1/2^p}{2^p - 1} k^p |\varphi_p(nk)| = \frac{k^p}{2^p} |\varphi_p(nk)| \le E.
$$

Thus, again neglecting terms of order k^{p+1}, we have, by (A.1),

$$
\big|v^{(2)}\big((2n)(k/2)\big) - y(nk)\big| \le E,
$$

in the same time interval in which the condition holds.

3.3 With $k = 0.1$ we obtain $T = 10.6$. The plot of both solutions $u^{(1)}$ (solid line) and $u^{(2)}$ (dashed line) are superimposed on the left. The plot on the right is the plot of $Q(t)$.

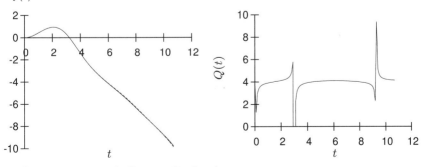

With $k = 0.01$ we obtain $T = 25.42$; the plots are

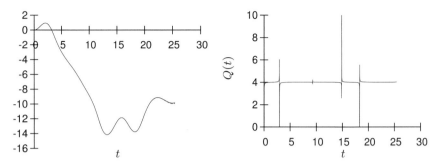

With $k = 0.001$ we obtain $T = 30.227$; the plots are

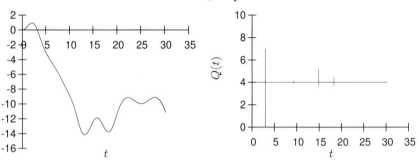

3.4 Plot of the solution:

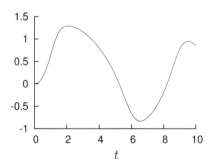

According to solution expansion (A.15), as the method is fourth-order accurate, the precision quotient $Q(t)$ satisfies

$$
\begin{aligned}
Q(t) &= \frac{v(t, k) - v(t, k/2)}{v(t, k/2) - v(t, k/4)} \\
&= \frac{y(t) + k^4 \varphi_4(t) + \mathcal{O}(k^5) - y(t) - (k/2)^4 \varphi_4(t) - \mathcal{O}(k^5)}{y(t) + (k/2)^4 \varphi_4(t) + \mathcal{O}(k^5) - y(t) - (k/4)^4 \varphi_4(t) - \mathcal{O}(k^5)} \\
&= \frac{1 - (1/2)^4 + \mathcal{O}(k)}{(1/2)^4 - (1/4)^4 + \mathcal{O}(k)} \\
&= 2^4 + \mathcal{O}(k).
\end{aligned}
$$

Thus, for k small enough, Q should approach the value 16. The plot of $Q(t)$ obtained is

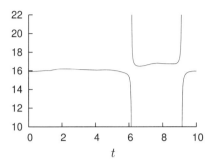

which shows that the code is working with the accuracy expected.

3.5 When applied to approximate the equation $dy/dt = \lambda y$, the improved Euler method reads

$$v_{n+1} = \left(1 + k\lambda + \tfrac{1}{2}k^2\lambda^2\right)v_n.$$

Then, with $\mu = k\lambda = x + iy$, the stability region in the complex μ-plane is given by

$$\sqrt{\left(1 + x + \tfrac{1}{2}(x^2 - y^2)\right)^2 + (y + xy)^2} \leq 1.$$

The plot of this stability region is

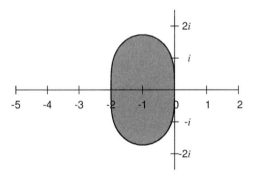

3.6 For the equation $dy/dt = i\lambda y$, as the time step $k > 0$, the stability region will be an interval in the imaginary axis. The third-order Taylor method becomes

$$v_{n+1} = \left(1 + ik\lambda - \frac{1}{2}k^2\lambda^2 - \frac{i}{6}\lambda^3 k^3\right)v_n.$$

Calling $\mu = \lambda k$, the stability interval in the imaginary axis is given by the condition

$$\left(1 - \tfrac{1}{2}\mu^2\right)^2 + \left(y - \tfrac{1}{6}\mu^3\right)^2 \leq 1,$$

which gives

$$\mu^4(\mu^2 - 3) \leq 0.$$

Therefore, the stability interval is $\mu \in [-\sqrt{3}, \sqrt{3}]$.

3.7 (a) With $\lambda = i\mu$, $\mu \in \mathbb{R}$, we have for the explicit Euler method

$$|1 + \lambda k| = \sqrt{1 + \mu^2 k^2} > 1 \quad \text{if } k \neq 0,$$

and onecan not choose k so that the method is stable.

(b) The modified Euler method proposed will be stable if

$$|1 + \lambda k + \alpha \lambda^2 k^2| \leq 1.$$

On the imaginary axis of the λk-plane, this condition is equivalent, writing $\lambda = i\mu$, to

$$1 - 2\alpha\mu^2 k^2 + \alpha^2 \mu^4 k^4 + \mu^2 k^2 \leq 1,$$

or, equivalently,

$$\alpha^2 (\mu k)^4 + (1 - 2\alpha)(\mu k)^2 \leq 0,$$

which holds when $(\mu k)^2 = 0$ (not interesting) or when $0 < (\mu k)^2 \leq (2\alpha - 1)/\alpha^2$. The right-hand side of the last inequality requires that $\alpha \geq \frac{1}{2}$ and reaches its maximum (maximum allowed interval for μk) when $\alpha = 1$.

3.8 Let $y(t)$ be the exact solution of $dy/dt = f(y,t)$, $y(0) = y_0$. Thus,

$$\frac{d^2y}{dt^2} = \frac{\partial f}{\partial y}(y,t)\frac{dy}{dt} + \frac{\partial f}{\partial t}(y,t) = \frac{\partial f}{\partial y}(y,t)f(y,t) + \frac{\partial f}{\partial t}(y,t), \tag{A.3}$$

where we have used the equation. Differentiating once more and omitting the dependence on (y,t), we get

$$\frac{d^3y}{dt^3} = \frac{\partial^2 f}{\partial y^2}f^2 + 2\frac{\partial^2 f}{\partial t \partial y}f + \left(\frac{\partial f}{\partial y}\right)^2 f + \frac{\partial f}{\partial y}\frac{\partial f}{\partial t} + \frac{\partial^2 f}{\partial t^2}. \tag{A.4}$$

(a) Now, by Definitions 3.2 and 3.7 we have, using Taylor expansions of y and f,

$$kR_n = y_{n+1} - y_n - kf\left(\frac{1}{2}\big(y_n + y_n + kf(y_n, t_n)\big), t_n + \frac{k}{2}\right)$$

$$= y_n + \left(\frac{dy}{dt}\right)_n k + \left(\frac{d^2y}{dt^2}\right)_n \frac{k^2}{2} + \left(\frac{d^3y}{dt^3}\right)_n \frac{k^3}{6} + \mathcal{O}(k^4) - y_n$$

$$- kf\left(y_n + \frac{k}{2}f_n, t_n + \frac{k}{2}\right)$$

$$= \left(\frac{dy}{dt}\right)_n k + \left(\frac{d^2y}{dt^2}\right)_n \frac{k^2}{2} + \left(\frac{d^3y}{dt^3}\right)_n \frac{k^3}{6} + \mathcal{O}(k^4) - k\left[f_n + \left(\frac{\partial f}{\partial y}\right)_n f_n \frac{k}{2}\right.$$

$$\left. + \left(\frac{\partial f}{\partial t}\right)_n \frac{k}{2} + \left(\frac{\partial^2 f}{\partial t^2}\right)_n f_n^2 \frac{k^2}{8} + \left(\frac{\partial^2 f}{\partial t \partial y}\right)_n f_n \frac{k^2}{4} + \left(\frac{\partial^2 f}{\partial t^2}\right)_n \frac{k^2}{8}\right]$$

$$+ \mathcal{O}(k^3).$$

In the expression above the terms independent of k cancel out by the differential equation, the linear terms in k cancel out by (A.3), and the cubic terms in k can be simplified by using (A.4), so that the truncation error turns out

$$R_n = \frac{k^2}{24}\left[\left(\frac{\partial^2 f}{\partial y^2}\right)_n f_n^2 + 2\left(\frac{\partial^2 f}{\partial t \partial y}\right)_n f_n + 4\left(\frac{\partial f}{\partial y}\right)_n^2 f_n + 4\left(\frac{\partial f}{\partial y}\right)_n\left(\frac{\partial f}{\partial t}\right)_n \right.$$
$$\left. + \left(\frac{\partial^2 f}{\partial t^2}\right)_n \right] + \mathcal{O}(k^3).$$

The method is accurate of order 2.

(b) For the method of Heun the analysis is very similar; we get

$$R_n = -\frac{k^2}{12}\left(\left(\frac{\partial^2 f}{\partial y^2}\right)_n f_n^2 + 2\left(\frac{\partial^2 f}{\partial t \partial y}\right)_n f_n + \left(\frac{\partial^2 f}{\partial t^2}\right)_n - 2\left(\frac{\partial f}{\partial y}\right)_n^2 f_n \right.$$
$$\left. - 2\left(\frac{\partial f}{\partial y}\right)_n\left(\frac{\partial f}{\partial t}\right)_n\right) + \mathcal{O}(k^3).$$

The method is accurate of order 2.

SOLUTIONS FOR CHAPTER 4

4.1 (b)

t_n	k_n	t_n	k_n
0.000000	6.666667×10^{-3}	7.262033×10^{-3}	7.707347×10^{-5}
0.000000	4.444444×10^{-3}	8.109841×10^{-3}	1.156102×10^{-4}
0.000000	2.962963×10^{-3}	8.919113×10^{-3}	1.734153×10^{-4}
0.000000	1.975309×10^{-3}	9.786189×10^{-3}	2.601229×10^{-4}
0.000000	1.316872×10^{-3}	1.082668×10^{-2}	3.901844×10^{-4}
0.000000	8.779150×10^{-4}	1.160705×10^{-2}	5.852766×10^{-4}
0.000000	5.852766×10^{-4}	1.277760×10^{-2}	8.779150×10^{-4}
0.000000	3.901844×10^{-4}	1.365552×10^{-2}	1.316872×10^{-3}
0.000000	2.601229×10^{-4}	1.497239×10^{-2}	1.975309×10^{-3}
0.000000	1.734153×10^{-4}	1.694770×10^{-2}	2.962963×10^{-3}
0.000000	1.156102×10^{-4}	1.991066×10^{-2}	4.444444×10^{-3}
0.000000	7.707347×10^{-5}	2.435511×10^{-2}	6.666667×10^{-3}
0.000000	5.138231×10^{-5}	3.102177×10^{-2}	1.000000×10^{-2}
0.000000	3.425487×10^{-5}	3.131022	1.500000×10^{-2}
6.439916×10^{-3}	5.138231×10^{-5}	3.146022	2.250000×10^{-2}

SOLUTIONS FOR CHAPTER 5

5.2 Proposing $v_n = \kappa^n$, equation (5.17) with $F(t) = 0$ implies that

$$(1 - k\eta)\kappa^2 - 2ik\xi - (1 + k\eta) = 0$$

whose solutions are, calling $x = k\eta$ and $y = k\xi$,

$$i\frac{y}{1 - x} \pm \frac{1}{1 - x}\sqrt{1 - x^2 - y^2}.$$

The method will be stable if there are two different roots with $|\kappa| \leq 1$. Thus, if $x^2 + y^2 < 1$, $x \leq 0$ the square root is nonzero and there are two different solutions; moreover,

$$|\kappa|^2 = \frac{1 - x^2}{(1 - x)^2} \leq 1.$$

5.3 The one-step Adams-Bashforth method for the equation $dy/dt = \lambda y$ is

$$v_{n+1} = v_n + k\beta_0 \lambda v_n.$$

By Taylor expansion of the exact solution

$$y_{n+1} = y_n + ky_{tn} + \mathcal{O}(k^2) = y_n + k\lambda y_n + \mathcal{O}(k^2).$$

Thus, the truncation error is

$$R_n = \frac{y_{n+1} - y_n}{k} - k\beta_0\lambda y_n = \lambda(y_n - \beta_0)y_n + \mathcal{O}(k).$$

It will be $\mathcal{O}(k)$ if one chooses $\beta = 1$, which gives the explicit Euler method. The two-step Adams-Bashforth method for the equation $dy/dy = \lambda y$ is

$$y_{n+1} = y_n + \lambda(\beta_0 y_n + \beta_1 y_{n-1}).$$

Using the Taylor expansions of the exact solution,

$$y_{n+1} = y_n + k\left(\frac{dy}{dt}\right)_n + \frac{k^2}{2}\left(\frac{d^2y}{dt^2}\right)_n + \mathcal{O}(k^3) = y_n + \lambda k y_n + \frac{\lambda^2 k}{2}y_n + \mathcal{O}(k^3),$$

$$y_{n-1} = y_n - k\left(\frac{dy}{dt}\right)_n + \frac{k^2}{2}\left(\frac{d^2y}{dt^2}\right)_n + \mathcal{O}(k^3) = y_n - \lambda k y_n + \frac{\lambda^2 k}{2}y_n + \mathcal{O}(k^3),$$

the truncation error becomes

$$R_n = \frac{y_{n+1} - y_n}{k} - \lambda(\beta_0 y_n + \beta_1 y_{n-1})$$

$$= y_n\left(\lambda(1 - \beta_0 - \beta_1) + \lambda^2 k\left(\tfrac{1}{2} + \beta_1\right)\right) + \mathcal{O}(k^2).$$

Therefore, we need $\beta_0 + \beta_1 = 1$ and $\beta_1 = -\frac{1}{2}$. The solution is $\beta_0 = \frac{3}{2}$ and $\beta_1 = -\frac{1}{2}$, and the two-step Adams-Bashforth for the general equation $dy/dt = f(y, t)$, is then

$$v_{n+1} = v_n + k\left(\frac{3}{2} f(v_n, t_n) - \frac{1}{2} f(v_{n-1}, t_{n-1})\right).$$

5.5 By continuity, it is enough to find the roots of $p_\mu(\kappa)$ and check their moduli for only one point in each region of interest. Calculating numerically, we get:

1. For $\mu = 0.3$, the roots are $0.015577 - i0.520760$, $0.015577 + i0.520760$, 0.307208, 1.349140. There is one root with module greater than 1, so the method is unstable in the outermost region.

2. For $\mu = 0.3 + i0.8$, the roots are $0.217037 - i0.480709$, $0.372203 + i0.046661$, $0.395738 + i1.234867$, $0.702522 + i1.032515$. There are two roots with module greater than 1, so the method is unstable in the region inside the upper loop.

3. For $\mu = 0.3 - i0.8$, the roots are the same as in the preceding case, so the method is also unstable inside the lower loop.

4. For $\mu = -0.1$, the roots are -0.523354, $0.194673 - i0.203202$, $0.194673 + i0.203202$, 0.904841. All the roots have module strictly smaller than 1. Thus, the method is stable inside the loop left of the origin.

SOLUTIONS FOR CHAPTER 7

7.4 (a) The Fourier coefficients are $\hat{s}(\omega) = (-1)^\omega i/(2\pi i)$.
(b) Plots for $M = 10$; truncated Fourier series on the left and error on the right:

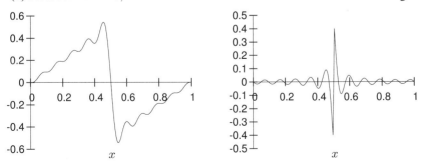

(c) Fourier interpolating polynomial on the left and error on the right:

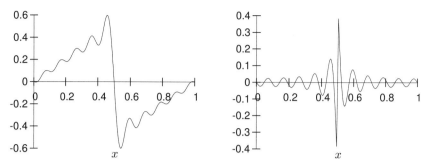

(d) Plots for $M = 100$; truncated Fourier series on the left and error on the right:

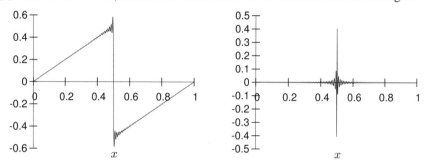

Fourier interpolating polynomial on the left and error on the right:

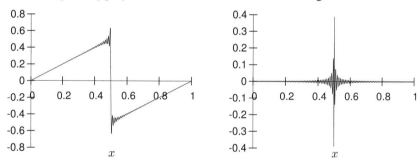

SOLUTIONS FOR CHAPTER 9

9.1 The approximation has to be antisymmetric around x. Then, writing it as

$$Df(x) = \frac{1}{h}(Af(x - 2h) + Bf(x - h) - Bf(x + h) - Af(x + 2h)),$$

using Taylor expansion around x and requiring that $Df(x)$ approximates $f'(x)$ to fourth order in h gives equations for A and B whose solutions are $A = \frac{1}{12}$ and $B = -\frac{2}{3}$. Thus, the approximation is

$$Df(x) = \frac{1}{12h}(f(x - 2h) - 8f(x - h) + 8f(x + h) - f(x + 2h)).$$

9.2 The span is five grid points and $\alpha = \frac{1}{12}$.

9.3 We have, by Taylor expansion,

$$
\begin{aligned}
D_+^2 D_-^2 v(x_j) &= \frac{1}{h^2}(v(x_j - 2h) - 4v(x_j - h) + 6v(x_j) \\
&\quad - 4v(x_j + h) + v(x_j + 2h)) \\
&= \frac{d^4 v}{dx^4}(x_j) + \frac{h^2}{6}\frac{d^6 v}{dx^6}(x_j) + \mathcal{O}(h^4).
\end{aligned}
$$

Therefore,

$$
(aD_0 - bh^2 D_+^2 D_-^2)v(x_j, t) + cv(x_j, t) = a\frac{\partial v}{\partial x}(x_j, t) + cv(x_j, t) + \mathcal{O}(h^2)
$$

and the approximation is second-order accurate.

9.5 (a) The exact solution is $u(x, t) = \sin(2\pi x)e^{-4\pi^2 t} + 10\sin(10\pi x)e^{-100\pi^2 t}$. The plots of the exact solution at times $t = 0$, $t = 0.004$, and $t = 0.2$ are

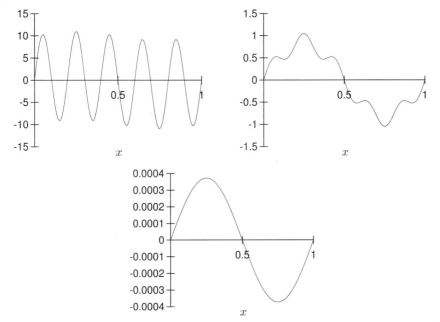

(b) and c) The plots of the solution (left plot) and error (right plot) are, for $h = 10^{-1}$ and $k = h^2/10$,

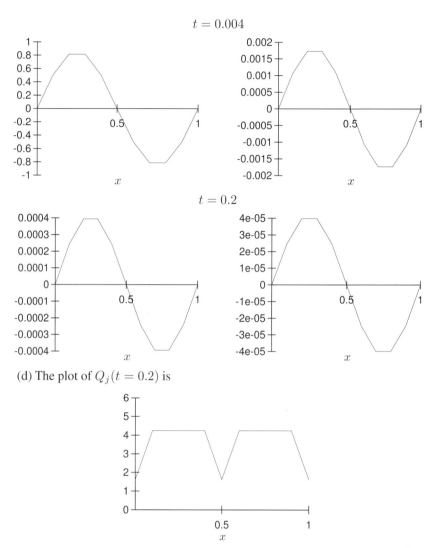

(d) The plot of $Q_j(t = 0.2)$ is

(e) The plots of the solution (left plot) and error (right plot) are, using $h = 10^{-2}$ and $k = h^2/10$,

$$t = 0.004$$

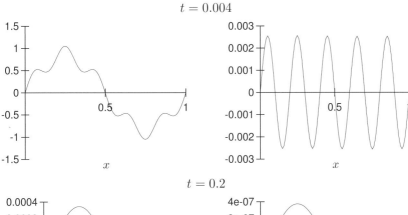

$$t = 0.2$$

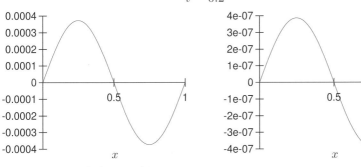

and the plot of $Q_j(t = 0.2)$ is

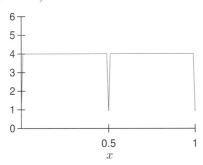

9.6 Calling $\mu = \lambda k^2$, the roots of (9.49) are

$$\kappa_\pm = \left(1 + \frac{\mu}{2}\right) \pm \sqrt{\mu + \frac{\mu^2}{4}}.$$

If $-4 < \mu < 0$, the discriminant is $\mu(1 + \mu/4) < 0$, and the square root is pure imaginary and different from zero. Thus, we have two different roots with

$$|\kappa_\pm|^2 = \left(1 + \frac{\mu}{2}\right)^2 + |\mu|\left(1 + \frac{\mu}{4}\right)$$

$$= 1 + \mu + \frac{\mu^2}{4} - \mu - \frac{\mu^2}{4} = 1.$$

SOLUTIONS FOR CHAPTER 10

10.2 (a) and (b) Plots for $t = 0.002$. Solution on the left and Q on the right.

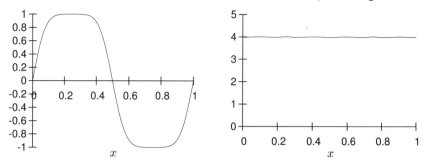

Plots for $t = 0.1$. Solution on the left and Q on the right.

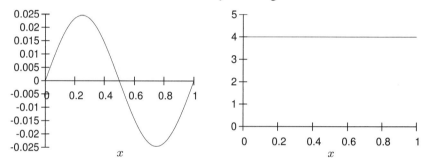

Plots for $t = 0.2$. Solution on the left and Q on the right.

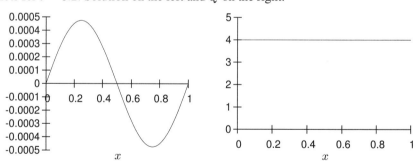

10.5 (a) The solution is C^1 smooth and constant along the characteristic lines $x+t = $ const. Thus,

$$u(x,t) = \begin{cases} 0 & \text{if } x+t < 1, \\ \frac{1}{2}(1 - \cos(2\pi(x+t-1))) & \text{if } 1 \leq x+t \leq 2. \\ 0 & \text{if } 2 < t. \end{cases}$$

(b) The plots of the numerical solution (left plots) and errors (right plots), at $t = 0.5$, $t = 1.0$, and $t = 1.5$ are, respectively

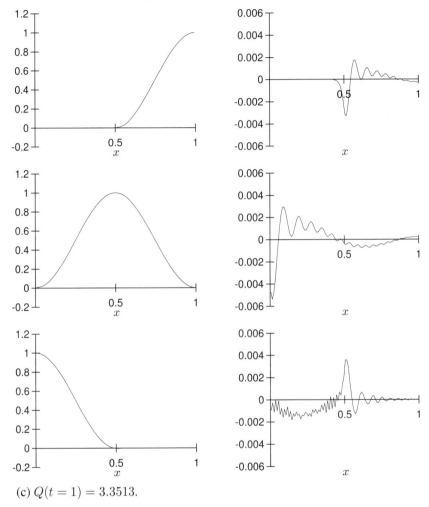

(c) $Q(t = 1) = 3.3513$.

REFERENCES

1. E. Oran Brigham. *The Fast Fourier Transform and Its Applications*. Prentice Hall, Englewood Cliffs, NJ, 1988.

2. H. S. Carslaw. *Introduction to the Theory of Fourier's Series and Integrals*. 2nd ed. Macmillan, London, 1921.

3. Earl A. Coddington. *An Introduction to Ordinary Differential Equations*. Dover, New York, 1989 (unabridged, corrected republication of the work first published by Prentice Hall, Englewood Cliffs, NJ, 1961).

4. Germund Dahlquist and Björck Ake. *Numerical Methods*. Dover, New York, 2003 (unabridged republication of the work first published by Prentice Hall, Englewood Cliffs, NJ, in 1974, which was an English translation of the work published as Numeriska Metoder by CWK, Gleerup, Lund, Sweden, in 1969).

5. Carl de Boor. *A Practical Guide to Splines*. (Rev. ed.) Springer-Verlag, New York, 2001.

6. Bertil Gustafsson, Heinz-Otto Kreiss, and Joseph Oliger. *Time Dependent Problems and Difference Methods*. Wiley-Interscience, New York, 1995.

7. Heinz-Otto Kreiss and Jens Lorenz. *Initial-Boundary Value Problems and the Navier-Stokes Equations*. SIAM, Philadelphia, 2004 (unabridged republication of the work first published by Academic Press, San Diego, CA, 1989).

8. Randall J. LeVeque. *Finite Difference Methods for Ordinary and Partial Differential Equations* (*Steady-State and Time Dependent Problems*). SIAM, Philadelphia, 2007.

9. Elias M. Stein and Rami Shakarchi. *Fourier Analysis* (*An Introduction*). Princeton University Press, Princeton and Oxford, 2003.

Introduction to Numerical Methods for Time Dependent Differential Equations, First Edition. **173**
By Heinz-O. Kreiss and Omar E. Ortiz. Copyright © 2014 John Wiley & Sons, Inc.

10. Georgi P. Tolstov. *Fourier Series*. Dover, New York, 1976 (unabridged republication, with slight corrections, of the work first published by Prentice-Hall, Englewood Cliffs, NJ, 1962).

Index